T0137044

Fluctuation-Induced Network Control and Learning

Masayuki Murata • Kenji Leibnitz

Editors

Fluctuation-Induced Network Control and Learning

Applying the Yuragi Principle of Brain
and Biological Systems

 Springer

Editors
Masayuki Murata
Graduate School of Information Science
and Technology
Osaka University
Suita, Osaka, Japan

Kenji Leibnitz
Center for Information and Neural
Networks
National Institute of Information and
Communications Technology (NICT)
Suita, Osaka, Japan

ISBN 978-981-33-4978-0   ISBN 978-981-33-4976-6   (eBook)
https://doi.org/10.1007/978-981-33-4976-6

This Springer imprint is published by the registered company Springer Nature Singapore Pte Ltd.
The registered company address is: 152 Beach Road, #21-01/04 Gateway East, Singapore 189721, Singapore

Preface

Future demands for information and communications technology will require new technologies that can operate much more efficiently and dynamically with its energy resources than current systems. To reach this goal, the design of new technologies could draw inspiration from the human brain that consumes only a fraction of the power that is used by a computer. For example, in 2016, Deep Mind's artificial intelligence (AI) system AlphaGo was able to beat the human world champion in the ancient board game of Go. While this achievement was a milestone indicating the great strides AI has taken over the recent years, researchers have also warned about the limits of deep learning due to its severe reliance on computationally intensive hardware. For example, AlphaGo required about 250 kW of electricity, while measurements have shown that the human brain consumes on average only about 20 W for operation. In addition, unlike AlphaGo that is only limited to the single task of playing Go, the human brain is able to simultaneously perform a wide variety of complex cognitive and motor tasks.

The secret of the brain's efficiency is believed to lie in the way how it utilizes "Yuragi," which is the Japanese term describing the noise or fluctuations that can be found in all biological systems including the human brain. Instead of computers utilizing vast amounts of input data and trying to eliminate noise to follow unambiguous rules, biological systems can actually take advantage of noise, which helps drastically to reduce the number of possible next actions with a high degree of freedom. When brain activity is actually measured, it undergoes large spontaneous fluctuations around various states in what is called the resting state (default mode network). We can consider this as the brain preparing possible attractors that represent candidate states that could be taken next. After fluctuating back and forth between these candidate states, one of these states is probabilistically selected as an appropriate attractor for the current situation. Whenever the environment situation changes, noise would drive the system state out of this attractor state toward a more suitable attractor. Such probabilistic behavior of the brain can also be described in terms of Bayesian statistics. By combining information from multiple modalities, such as visual or auditory information, the brain must decide on the next actions to take according to the uncertainty of the information. The perceptual model of the Bayesian attractor model (BAM) can handle three kinds of such uncertainties, which can be utilized to perform an adequate control in the presence of noisy input information.

In this book, we summarize key research work that is based on this Yuragi concept utilizing attractors in the presence of noisy dynamics. This book consists of two parts. Part I includes four chapters that are the direct result of interdisciplinary work performed at various research groups at Osaka University, Japan, between 2004 and 2008. Part II includes additional contributions that extend the original Yuragi concept to a BAM and presents applications to various fields in information network control and AI. With the growing complexities of mechanisms that are nowadays involved in network protocols, we believe that Yuragi control provides a simple, yet effective means of dynamically controlling large-scale systems and networks by following ideas that can be observed from models in biology and neuroscience. With predictions that its future demands for computing power will impose a limit on how far deep learning can improve in performance, progress in machine learning is expected to move more toward less computationally intense methods, such as Yuragi-based learning. While it may still be not fully proven whether the brain is in fact operating in a probabilistic manner or not, its ability to handle uncertainty can nevertheless be used as inspiration to develop simple and efficient control mechanisms in dynamically changing systems.

Finally, we would like to take this opportunity to thank all authors who supported this book with their contributions.

Osaka, Japan   Masayuki Murata
Osaka, Japan   Kenji Leibnitz
September 2020

Contents

Contributors

Onur Alparslan Graduate School of Information Science and Technology, Osaka University, Suita, Osaka, Japan

Shin'ichi Arakawa Graduate School of Information Science and Technology, Osaka University, Suita, Osaka, Japan

Toshiyuki Kanoh Industry-Academia Collaboration, Graduate School of Information Science and Technology, NEC Brain Inspired Computing Alliance Laboratories, Osaka University, Suita, Osaka, Japan

Yuki Koizumi Graduate School of Information Science and Technology, Osaka University, Suita, Osaka, Japan

Daichi Kominami Graduate School of Information Science and Technology, Osaka University, Suita, Osaka, Japan

Kenji Leibnitz Center for Information and Neural Networks, National Institute of Information and Communications Technology, Suita, Osaka, Japan

Masayuki Murata Graduate School of Information Science and Technology, Osaka University, Suita, Osaka, Japan

Tsutomu Murata Center for Information and Neural Networks, National Institute of Information and Communications Technology, Suita, Osaka, Japan

Yuichi Ohsita Graduate School of Information Science and Technology, Osaka University, Suita, Osaka, Japan

Tatsuya Otoshi Graduate School of Economics, Osaka University, Toyonaka, Osaka, Japan

Naoki Wakamiya Graduate School of Information Science and Technology, Osaka University, Suita, Osaka, Japan

Toshio Yanagida Center for Information and Neural Networks, National Institute of Information and Communications Technology, Suita, Osaka, Japan

Part I
Fluctuation-Based Control Systems:
Yuragi Concept

Chapter 1
Introduction to Yuragi Theory and Yuragi Control

Kenji Leibnitz

Abstract Noise and fluctuations are phenomena that are frequently observed in biology and nature, but also intrinsically occur in various types of technological and engineered systems. In this chapter, we provide an introduction to the concepts and methods that underlie Yuragi-based control mechanisms. *Yuragi* is the Japanese term for fluctuations, and this concept can be utilized for simple yet effective control mechanisms to adaptively control information and communication systems depending on the environment with simple rules. In this chapter, we present several examples to illustrate how fluctuations occur in biological systems and how stochastic biological models can be utilized to design new robust and flexible control algorithms, such as attractor selection and attractor perturbation mechanisms.

1.1 Introduction

During transmission over any type of communication medium, noise or fluctuations eventually appear as variations in the received signal level or transmission speed. This is caused by a multitude of natural and artificial factors that influence the quality of the wired or wireless communication channel, as well as by the non-deterministic behavior of the traffic sources.

The most basic form of noise is thermal noise (Johnson–Nyquist noise), which is generated by the random thermal motion of electrons at the receiver and depends on the temperature of the environment. The received amplitude of this noise resembles a Gaussian probability density function; therefore, communication channels are often modeled as additive white Gaussian noise channels. Another type of noise is shot noise [32], which is caused by fluctuations of the electric current due to the discrete nature of the arrival of electrons in the devices. Flicker noise ($1/f$ noise or pink noise) has a frequency spectrum that is inversely proportional to the frequency

K. Leibnitz (✉)
Center for Information and Neural Networks, National Institute of Information and
Communications Technology, Suita, Osaka, Japan
e-mail: leibnitz@nict.go.jp

© Springer Nature Singapore Pte Ltd. 2021
M. Murata, K. Leibnitz (eds.), *Fluctuation-Induced Network Control and Learning*,
https://doi.org/10.1007/978-981-33-4976-6_1

of the signal, while burst noise consists of sudden step-like transitions between two or more discrete voltage or current levels, resulting in the crackling sounds often experienced in audio circuits. Finally, transit-time noise reflects the time it takes for the current to travel from the input to the output and appears as a random noise proportional to the frequency of operation.

Due to the various types of noise as well as the presence of nearby transmitters with interfering traffic, a key metric for evaluating the received signal quality over a communication channel is the signal-to-noise ratio (SNR), which represents the proportion of the level of the desired signal to the level of noise. By designing communication protocols that can maintain a high SNR, systems can achieve a high transmission speed as well as high channel utilization.

Following this traditional engineering viewpoint, it is generally considered that noise is detrimental to communication quality and should be eliminated as much as possible. However, there are cases in which utilizing noise is beneficial for the system. For example, in code division multiple access (CDMA) [62], noise-like random codes are used as carriers for modulating individual user signals. CDMA-based systems have proven to be very efficient by operating close to the Shannon capacity of the communication channel.

In biological systems, noise can also be observed in various forms. On a microscopic level, for instance, cells may be of the same type; however, the quantities describing them or their concentrations vary among cells and fluctuate over time [25]. The entire process of signal transduction within a cell is driven by diffusion and fluctuations. On a macroscopic level, the dynamics of populations of species and their entire genetic evolution is influenced by noise, such as the random mutation of genes in DNA. Thus, noise is a fundamental component that is inherent to many types of biological processes [68], and it is of great importance for the adaptability and robustness of biological systems.

By algorithmically mimicking the behavior of noise-driven biological mechanisms, researchers have developed probabilistic solutions of NP-hard optimization problems with heuristics such as genetic algorithms (GAs) [23] and simulated annealing [1]. These noise-driven algorithms share the property that they generally do not operate on fixed rules in searching for solutions; instead, they utilize random search to find adequate solutions in a significantly shorter time.

In the remainder of this chapter, we discuss how noise can be utilized to find solutions that solve problems in a simple and efficient way. We refer to these methods as Yuragi methods, using the attractor selection method as a key example. In Sect. 1.2, we first discuss the fundamentals of self-organization and the key role that noise plays. In Sect. 1.3, we present examples in which noise is essential for characterizing and solving problems. In Sect. 1.4, we provide a mathematical formulation of noise-driven systems, and review the concept of attractors and stability. In Sect. 1.5, we define the basic Yuragi model for attractor selection and present three examples of how attractors can be formulated. Finally, in Sect. 1.6, we conclude this chapter with a discussion of the benefits of utilizing noise in information networking.

1.2 Principles of Self-organization

Self-organization is a key feature of biological systems. In this section, we briefly describe the difference between centralized and distributed control, and summarize the characteristic features of self-organized systems. We then discuss how utilizing noise can contribute to self-organization.

1.2.1 Centralized and Distributed Control

Networking protocols were originally designed to operate in a client/server structure, in which a single server controls access to resources, and to access these resources, each client must be connected to the server. Because the central server has access to the state of the entire system, it can assign resources in a globally optimal way. However, due to the finite capacity of the server, this method quickly reaches a limit when the number of accessing client nodes dramatically increases.

This limitation can be alleviated by introducing a distributed type of control of the resources. In self-organized network protocols, such as peer-to-peer networks, each node simultaneously acts as a server and client and manages the available resources among its peer nodes without requiring a central entity. Each node makes its own control decisions based entirely on locally available information. This can lead to inferior results in performance compared to centralized control; however, the mechanism becomes scalable with the number of participating nodes, and more robust when faced with unforeseen events. Biological systems generally operate in a self-organized manner without having explicit centralized control, and each individual contributes to the group. Such behavior is referred to as swarm intelligence [8] and can be observed in the collaboration of social insects, such as ant foraging.

Figure 1.1 illustrates the relationship between scalability and controllability depending on the types of control [15]. The left part of the figure displays a system with centralized control (e.g., a traditional star-shaped client/server system), while the right part of the figure displays a fully self-organized system (e.g., a peer-to-peer network). Distributed systems can be considered intermediate structures that are combinations of both architectures, thereby allowing for the management of a large number of nodes in a more scalable way while preserving the benefits of centralized control.

It should be noted that security can become a concern when dealing, for instance, with content distribution systems. Here, content provided by centralized servers can be easily authenticated and verified, whereas in distributed and self-organized systems, methods must be implemented to secure the content from tampering [35].

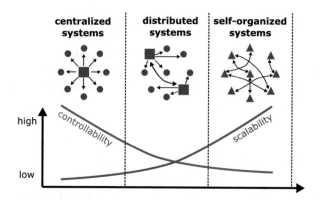

Fig. 1.1 Trade-off between centralized, distributed, and self-organized systems in terms of controllability and scalability

1.2.2 Characteristics of Self-organized Systems

Self-organization occurs in systems consisting of autonomous agents, such as cells or insects that collaborate in various structures without utilizing explicit rules. Self-organization is closely related to the emergent property of biological systems [8], in which the outcome of the system depends on the collection of individual behaviors and their interactions, rather than each individual task performed.

In general, four fundamental features can be observed in the self-organization of biological systems: positive feedback, negative feedback, distributed and autonomous operation, and randomness and fluctuation. Positive feedback (e.g., recruitment or reinforcement) permits the system to evolve and promotes the creation of structure. Positive feedback serves as an amplifier for a desired outcome, whereas negative feedback regulates the influence from previous bad adaptations. Negative feedback prevents the system from becoming trapped in local solutions, and may take the form of saturation, exhaustion, or competition. Nature-inspired systems also tend not to rely on a global control unit, but operate in an entirely distributed and autonomous manner. This signifies that each agent acquires information, processes it, and stores it locally. However, to generate a self-organized structure, agents must exchange information with each other, which is achieved either by direct or indirect interaction. Another characteristic commonly observed in self-organized structures is that these structures often rely on randomness and fluctuation to enable the discovery of new solutions and enhance the stability and resilience of the system.

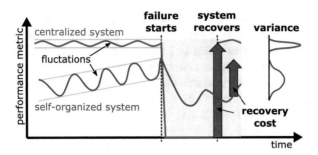

Fig. 1.2 Comparison of the recovery of a centralized and self-organized system after a critical error. Whereas the centralized system operates at a higher performance level than the self-organized system under normal conditions, it suffers more severely when critical errors occur and often requires manual intervention to recover

1.2.3   Role of Noise in Self-organized Systems

Whereas most engineered systems are designed to function optimally under certain controlled operational conditions, biologically inspired systems utilize fluctuations and randomness in their adaptations and can thus better cope with changes in their environment. This can be observed in the systems' reactions to critical error, as illustrated in Fig. 1.2. When a centralized system encounters a critical error due, for example, to the failure of a bottleneck server or a drastic change in operational conditions, its performance will decrease until the failure is manually corrected.

In contrast, self-organized systems are less dependent on individual network elements, because the system output emerges from a collaboration between units. Thus, a self-organized system is able to maintain an adequate operational level. However, because each element in the system is not aware of the global objective of the system, each node may only have a small contribution to the overall recovery, signifying that many adaptation steps may be necessary to reach an acceptable level of performance after failure.

Figure 1.2 also reveals that the variance of the performance metric likely differs between centralized and self-organized systems. The amount of fluctuation can therefore be regarded as being indirectly proportional to the cost necessary to return the performance to an adequate level. This effect is discussed in greater detail in Sect. 1.4.3.

1.3   Examples of Nature-Inspired Models Utilizing Noise

The cumulative dynamic behavior of any biological or technical system is often heavily influenced by an inherent stochastic component. The natural shapes of individual units are never perfect, nor are their actions truly deterministic, which leads to variations in the shape and behavior of biological cells [25]. In this section, we examine several examples of fluctuations being utilized in natural or nature-inspired systems to drive the systems toward certain states. In their simplest form, these fluctuations can be modeled as a stochastic process in which the numerical values of a system metric vary randomly over time.

We now present a brief introduction to several well-known mathematical models of biological mechanisms that make use of system-inherent noise to control and optimize the system. Our focus is mainly on the models themselves. As a result, we provide only a brief description, and the interested reader is referred to the literature for further details. It should be noted that the following examples are not intended to be exhaustive, but instead are limited to well-studied models that have been applied to information science in the past.

1.3.1   Random Walks and Brownian Motion

A random walk describes a path of successive random steps performed in some form of mathematical space. This makes it a special case of a Markov process. In biology or physics, random walks are often observed in the form of Brownian motion, which describes the random motion of a particle immersed in fluid. In mathematics, such a process is defined as a Wiener process. A particle undergoes continuous-time random motion due to collisions with the molecules of the fluid, and its spatial displacement over time is characterized by the Langevin equation, which is expressed in Eq. (1.1).

$$\frac{dv(t)}{dt} = -\gamma \, v(t) + \eta(t). \tag{1.1}$$

In Eq. (1.1), $v(t)$ is the velocity of the Brownian particle as a function of time t, γ is the friction coefficient of the underlying fluid, and $\eta(t)$ is a Gaussian random noise term that characterizes the fluctuations inherent in the system. The Langevin equation can be solved by the Fokker–Planck equation to obtain the temporal evolution of the probability density function of the position and velocity of the particle. An example of Brownian motion is depicted in Fig. 1.3a.

In Brownian motion, the distribution of diffused distances follows a bell-shaped curve. In contrast, in a Lévy flight, these distances follow a power law, which results in a pattern with many short steps that are occasionally interrupted by long steps spanning multiple orders of magnitude. Such distributions have been observed in real-world data of the foraging patterns of numerous marine animals [58] and the human mobility patterns of mobile phone users [21].

It should be noted that while Brownian motion or Lévy flight is performed over a continuous state space, a random walk can also be performed on discrete structures, such as along the edges of graphs [44]. Here, as illustrated in Fig. 1.3b, a random walk is performed from one node to one of its neighbors, where the probability of selecting a link is proportional to the node's outgoing degree. Additional variants

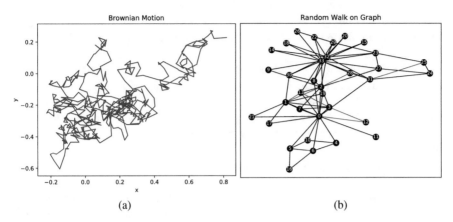

Fig. 1.3 Simulated trajectories of (**a**) Brownian motion (1000 steps) and (**b**) a random walk on a graph (10 steps)

of random walks exist, such as the maximum entropy random walk, in which the selection probabilities of the next-hop node are biased according to the principle of maximum entropy, leading to a more uniform visiting probability distribution among all paths in the graph [10, 59].

A random walk on a graph can also be used as a metric for characterizing the importance of individual nodes within the graph structure. For example, consider an agent that continuously hops from one node to one of its neighbors at random. Depending on the graph's connectivity structure, this agent will visit nodes that have many incoming links from other nodes more frequently. This is the underlying mechanism behind the *PageRank* [9] algorithm, which determines the order of importance of nodes in a graph and has become the basis of modern Internet search engines.

1.3.2   *Impact of Noise on Visual Perception and Decision-Making in the Brain*

The relationship between noise and neural signals in the human brain is multi-faceted. In general, the influence of physiological noise and other confounds is considered detrimental [6]. For instance, it is common to remove the contributions of heartbeat and respiration from recorded signals after acquisition. Other variables derived from the data are also usually corrected, such as the average signal fluctuations in white matter and cerebrospinal fluid.

In addition to these imperfections in recording, noise can also be caused by the underlying biophysical mechanisms of neurons [66], such as responses to repeated stimuli, which differ per trial due to the randomness in the opening and closing of ion channels at neural synapses. Noise can also emerge in the synaptic process due to multiple causes, such as the random nature of diffusion and chemical reactions within the synaptic cleft, and the unpredictable responses of ligand-gated ion channels.

In the literature, it is speculated that noise may serve a number of purposes in the nervous system [66]. For example, noise can expand the dynamic behavior of neurons as well as assist in enhancing signal detection. Ermentrout et al. [16] suggested that aspects of neuronal responses that appear as noise may in fact be a fundamental component in which information is propagated or represented in neurons.

On a more macroscopic level, diffusion models have been used to characterize how noise influences neural population dynamics involved in the process of decision-making. In [11], it was demonstrated that diffusion models are well suited for describing the perceptual performance of an experimental subject and the resulting reaction time required to make a decision. In these models, the information from a stimulus is represented by a differential equation with the mean drift rate of a random variable, which is then accumulated over time until reaching an upper or lower boundary, as illustrated in Fig. 1.4. The time required to cross one of the decision boundaries reflects the process of making a specific decision. These boundaries can also be adjusted to reflect the speed of decision-making.

A similar model has also been applied to a process called emergent recognition of partially degraded images without possessing any prior top-down information [50] (see also Chap. 2). Experiments with functional magnetic resonance imaging and behavioral tests verified that such a diffusion model is consistent with the human process of recognizing an image depending on two parameters: the difficulty of the image and the capability of the subject.

In a different experiment, multi-stable states were used for modeling bistable perception and binocular rivalry. Binocular rivalry occurs when one image is presented

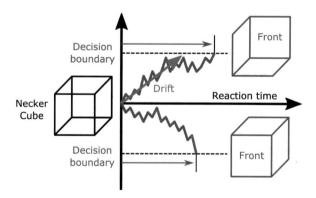

Fig. 1.4 Sketch of drift diffusion model for characterizing the orientation of a Necker cube with two simulated diffusion paths. The time needed for recognizing whether the cube is facing to top right (blue) or bottom left (red) corresponds to the time of the diffusion process crossing either of the decision boundaries

to a person's left eye and a different image to the right eye, leading the person to alternate in conscious perception [7]. Perceptions can change spontaneously from one interpretation to another even when the visual input is constant. One example is the Necker cube, as illustrated in Fig. 1.4, in which visual perception occasionally flips to make a different edge of the cube appear closer to the observer.

Neuronal fluctuations can thus be advantageous to brain function because they lead to a probabilistic behavior in decision-making by preventing deadlocks and are also important in signal detection. The presence of neuronal or synaptic noise makes neurons more sensitive to a broader range of inputs and improves the temporal and spatial discrimination of input signals.

1.3.3 Signal Enhancement Through Stochastic Resonance

As demonstrated in the examples presented in Sect. 1.3.2, all biological systems must handle noise that is inherent in the environment. Stochastic resonance occurs when the SNR of a nonlinear system increases through the addition of a moderate amount of noise [20, 47, 65, 67]. The frequencies of the white noise that correspond to the original signal resonate and amplify the original signal, resulting in an increase in the SNR. This effect can be observed in nature, for instance, when crawfish (*Procambarus clarkii*) make use of fluctuations in the water flow to detect approaching fish through small and inconspicuous movements [48].

Consider the bistable potential illustrated in Fig. 1.5, in which a particle exists in one of two basins [63]. The potential changes between the two different states are indicated by different colors. Now, consider that we introduce a certain amount

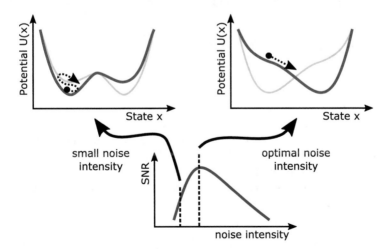

Fig. 1.5 Influence of noise intensity on stochastic resonance dynamics. The particle can cross the barrier only if the noise intensity is in the optimal range

of noise as the source of perturbation in the potential. If the noise intensity is too small, the particle cannot cross the barrier and remains trapped in its basin (top left of Fig. 1.5). However, if the noise is sufficiently strong, the bottom of the basin with the particle rises above the barrier, and the particle can easily move into the other basin (top right of Fig. 1.5). With an appropriate level of noise intensity, the probability that the particle can move across basins is highest when the source signal is strongest. However, if the noise intensity is too large, the movement of the particle becomes completely independent of the weak original signal. This implies that the correlation, or SNR, between the particle movement and weak original signal has a peak at a certain noise intensity (bottom of Fig. 1.5).

These dynamics can be formulated by the following first-order Langevin equation

$$\frac{dx(t)}{dt} = -\frac{dU(x(t))}{dx} + \epsilon \sin(\omega t) + \eta(t), \tag{1.2}$$

where $x(t)$ is the position of the particle at time t, function $U(x)$ defines the base potential, $\epsilon \sin(\omega t)$ corresponds to the weak original signal, and $\eta(t)$ is white Gaussian noise. An example of the potential is $U(x) = -\alpha x^2 + \beta x^4$, where $\alpha = 1/2$ and $\beta = 1/4$. For a bistable potential model, the output signal has only two different ranges of values, namely the left and right basins that can be discretized by placing a threshold at the center. The transition between the two states is well synchronized with the original signal at double the frequency when the noise intensity is at an appropriate level.

A network of multi-threshold devices also stochastic resonance behavior even for suprathreshold signals, whereas with a single threshold, stochastic resonance is observed only for subthreshold signals [46, 60]. For the input signal $x(t)$ containing noise, the output signal $y(t)$ of a multi-threshold network is defined as

$$y(t) = \sum_{i=1}^{N} y_i(t) \quad \text{with} \quad y_i(t) = \begin{cases} 1, & \text{if } x(t) + \eta_i(t) > \theta \\ 0, & \text{otherwise} \end{cases}, \tag{1.3}$$

where N is the number of threshold devices, $\eta_i(t)$ is the local noise at the threshold device i, and θ_i is its threshold.

There have been various applications of stochastic resonance. For example, Mizutani et al. [49] used stochastic resonance to consider the noise-assisted detection of binary signals in a distributed sensor network. Here, noisy signals from multiple binary detectors are combined at a single data fusion center. Other applications used stochastic resonance to enhance processing in both theoretical models of neural systems and experimental neuroscience [45].

1.3.4   Evolutionary and Genetic Algorithms

As discussed in Sect. 1.1, biological systems often rely on feedback-based control loops to adapt to changes in the environment. Such phenomena are also observed in the natural selection process of evolution. Species evolve over many generations to better adapt to a dynamically changing habitat by modifying their genetic building blocks. Mutations introduce diversity through generations to increase adaptability to new challenges.

Genetic algorithms (GAs) are a popular class of evolutionary algorithms that are commonly used as a heuristic to solve optimization problems. In GAs, candidate solutions of a problem are encoded as individuals (genomes), and evolution is performed over several generations where operations of natural selection, mutation, and reproduction are applied to these individuals. Selection is the process in which the fitness of individuals is evaluated, upon which the new generation is selected. From the pool of individuals, parents are selected, which then produce offspring by crossover and mutation, as illustrated in Fig. 1.6a. In crossover, parts of the chromosomes of the parents are split and recombined to form new offspring chromosomes. Mutation further increases the diversity by randomly modifying the genetic material. This process is repeated over several generations until a termination condition is reached. The resulting solution corresponds to an individual with the best adaptation to its environment.

GAs are commonly used as a heuristic to determine solutions for problems that remain static and have a single fitness function. However, when the environment changes over time, the system may need to reflect this property by also adapting its goals, and thus, its fitness. By dynamically alternating between two or more varying objectives, the system can evolve toward intermediate states that are highly adaptable by easily reaching all of the goals. Such a meta-algorithm, known as modularly varying goals (MVG), was proposed by Kashtan et al. [28] and is illustrated in Fig. 1.6b. It controls an underlying GA so that it can more easily adapt to multiple varying environments, which can also help accelerate evolution [29]. In an application for placing network functions in software-defined networks, modified versions of MVG were used in [54, 55].

Fig. 1.6 (**a**) Standard GA, in which mutations and crossovers shift the genetic population toward the highest peak in the fitness landscape. (**b**) MVG, in which the system iterates between two different fitness goals to reach a solution that is midway between both

1.3.5 Routing Methods Inspired by Social Insects

Another example in nature where randomness in biological dynamics is utilized to solve problems in computer science is the ant colony optimization (ACO) meta-heuristic introduced by Dorigo et al. [14]. ACO can provide near-optimal solutions for finding shortest path routes between source and destination nodes in a graph. This is an essential problem in networking, and many variants of ACO have been applied to routing in scenarios with stationary and mobile nodes [17]. Some of these variants are introduced in the Sects. 1.3.5.1–1.3.5.4, and additional information can be found in [2].

1.3.5.1 Ant Colony Optimization (ACO)

ACO is a probabilistic optimization scheme for finding the shortest path in a graph. It is based on the foraging behavior of ants finding their way from a food source to their nest. ACO can thus find a solution to the traveling salesman problem, and its strength lies in its ability to dynamically adapt to temporal changes in the graph topology, when links are added or removed.

The fundamental mechanism of ACO is illustrated in Fig. 1.7. Artificial ants (agents) wander randomly along the graph in search of food and return to their nest once they find a food source. Along their paths, the ants lay trails of chemical pheromones that act as orientation landmarks to indicate to themselves and other ants the paths that were taken. As these pheromones evaporate over time, inefficient paths that are not refreshed are discarded following the principle of negative feedback described in Sect. 1.2. Positive feedback, in contrast, rewards successful paths leading to the food source, as these paths are reinforced by other ants that follow the same trail and increase the pheromone density.

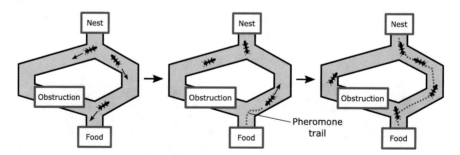

Fig. 1.7 In ant colony optimization, new paths are first randomly selected by ants. When a successful path is found, an ant lays a trail of pheromones on its return trip to the nest. Other ants finding this trail will also follow it, strengthening it over time. As pheromones evaporate over time, the probability of selecting inefficient paths decreases

1.3.5.2   AntNet in Packet-Switched Networks

AntNet, an algorithm proposed by Di Caro and Dorigo [12], is an application of ACO to routing in packet-switched networks. The operation of AntNet is performed by two different types of ants. At regular intervals, each router in the network sends forward ants toward randomly selected destinations to probe the network for a minimal-cost route. Whenever forward ants encounter a router on their way, they randomly select the next hop depending on the probabilities in the router table. If the destination node is unknown at a router, the next hop is selected uniformly at random among all possible candidates. Once the forward ants reach their destination, they transform into backward ants and follow the same reverse path back to the source node. During their return trip, backward ants update the entries of the routing tables at each intermediate router according to a metric that evaluates the quality of the path. When a backward ant reaches its original node, it dies and is removed from the system.

1.3.5.3   AntHocNet in Mobile Ad-Hoc Networks

Due to its capability of operating in a changing environment, ACO also provides an effective solution for routing in mobile ad-hoc networks. *AntHocNet* [13] is built on ACO routing and uses a hybrid approach to multipath routing. Whereas AntNet uses a proactive scheme by periodically generating ants for all possible destinations, AntHocNet consists of both proactive and reactive components. Paths are only established to destinations when they are needed by reactive forward ants that are launched by the source node to find multiple paths to the destination. Backward ants return to the source to establish the paths. Data are routed stochastically over different paths stored in the pheromone tables, and the tables are continuously updated by proactive forward ants. The algorithm reacts to link failures either with a local route repair or by warning preceding nodes along the paths. In [13], the authors demonstrated that AntHocNet can outperform ad-hoc on-demand distance vector (AODV) routing in terms of performance and scalability.

1.3.5.4   BeeHive for Wired Connectionless Networks

The social behavior of not only ants, but other social insects such as bees, has been modeled for their swarm behavior in routing [64]. Similar to AntNet, in Beehive, new paths are continuously examined to discover and update routes according to the current network conditions. BeeHive uses two different types of bee agents: one for covering short distances and the other for covering long distances within the network. Both types of bees are generated at the nodes and perform the same task of exploration of the network and evaluation of the traversed paths to update the routing tables. Their only distinction is in the distances they cover in the network.

The network is subdivided into foraging zones and foraging regions. A foraging zone is the local neighborhood of a node in which short-distance bee agents can reach the node. Thus, any node can simultaneously belong to multiple foraging zones of different nodes. From a global perspective, the network is also a collection of non-overlapping clusters (foraging regions) to which each node exclusively belongs. Each foraging region has a single representative node from which long-distance bee agents are launched. Each node maintains the routing information for all nodes within its foraging zone as well as for representative nodes of different foraging regions. If the destination of a packet lies outside the foraging zone of a node, it is forwarded to the representative node of the foraging region containing the destination node.

1.3.6 Tug-of-War Model for Solving the Multi-Armed Bandit Problem

In [51], Nakagaki et al. experimentally demonstrated that slime mold *Physarum polycephalum* can be used to empirically find the shortest path in a maze. The slime mold maintains a constant volume of the entire cell while collecting environmental information by concurrently expanding and shrinking its branches utilizing intrinsic fluctuations. This behavior was formulated in [3, 30] as a concurrent decision-making scheme for solving the multi-armed bandit problem (MABP). In the MABP, a player faces N different slot machines in an exploration–exploitation dilemma to select the machines that the player wishes to play. The dynamics and expected payout of each slot machine are unknown a priori, and the goal of the player is to select a sequence of machines that will maximize the winnings. The authors formulated a solution to this problem that they called *tug-of-war*, and additional studies in [31, 52] demonstrated its applicability in a physical implementation with quantum dots and optical near-field interactions. The tug-of-war concept has also been applied in the field of communication networks as an adaptive channel selection scheme in Internet of Things (IoT) environments [34, 43].

1.4 Mathematical Formulation of Noise-Driven Systems

In this section, we follow the notations in [41] to provide a formal introduction to dynamic systems before presenting the mathematical definitions of the Yuragi model and introducing its extensions. We provide only the basics in this section, and the reader is referred to well-known mathematics textbooks, such as [61], for further details.

1.4.1 Stability and Attractors

In the following, we consider an arbitrary, time-dependent N-dimensional system with $N > 1$, where time will be denoted t. The system state is described as a vector $\mathbf{x}(t)$ as follows:

$$\mathbf{x}(t) = \begin{pmatrix} x_1(t) \\ \vdots \\ x_N(t) \end{pmatrix},$$

where we assume that $x_i(t)$ is well defined in a real-valued, non-negative domain $\mathcal{D} \subset \mathbb{R}_0^+$ for all $i = 1, \ldots, N$. Note that for the sake of simplicity, we only indicate the variable denoting time t if it is explicitly required.

The dynamic behavior of this system is described by its derivatives over time t. We formulate this through a vectorized function $\mathbf{F}(\mathbf{x})$ in a differential equation system, as given in Eq. (1.4):

$$\frac{d\mathbf{x}}{dt} = \mathbf{F}(\mathbf{x}) = \begin{pmatrix} f_1(\mathbf{x}) \\ \vdots \\ f_N(\mathbf{x}) \end{pmatrix}, \tag{1.4}$$

where functions $f_i : \mathcal{D}^N \to \mathcal{D}, i = 1, \ldots, N$ determine the dynamic behavior in each dimension of the system. When we observe the system state asymptotically over time, we can formulate the steady state vectors $\mathbf{x}^*$ as solutions of Eq. (1.5) when there is no change over time; that is,

$$\frac{d\mathbf{x}}{dt} = \mathbf{0}. \tag{1.5}$$

We call a solution of Eq. (1.5) the *equilibrium point* $\mathbf{x}^*$ of the system. Depending on the structure of the functions $f_i(\mathbf{x})$, one or more equilibrium point may exist in the entire state space. The following formal definitions of stability of an equilibrium point $\mathbf{x}^*$ are generally used [61] as illustrated in Fig. 1.8.

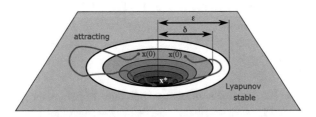

Fig. 1.8 Definitions of stability of $\mathbf{x}^*$ for a dynamic trajectory starting at $\mathbf{x}(0)$

Definition 1.1 (Stability) Let $\mathbf{x}^*$ be an equilibrium point of a system, as given in Eq. (1.4). Then, $\mathbf{x}^*$ is defined as *attracting* if

$$\exists_{\delta>0} : \|\mathbf{x}(0) - \mathbf{x}^*\| < \delta \Rightarrow \lim_{t\to\infty} \mathbf{x}(t) = \mathbf{x}^*.$$

We state that $\mathbf{x}^*$ is *Lyapunov stable* if

$$\forall_{\varepsilon>0} \, \exists_{\delta>0} : \|\mathbf{x}(0) - \mathbf{x}^*\| < \delta \Rightarrow \forall_{t\geq 0} : \|\mathbf{x}(t) - \mathbf{x}^*\| < \varepsilon.$$

If $\mathbf{x}^*$ is both attracting and Lyapunov stable, it is called *asymptotically stable*. Finally, point $\mathbf{x}^*$ is *exponentially stable* if

$$\exists_{\alpha,\beta,\delta>0} : \|\mathbf{x}(0) - \mathbf{x}^*\| < \delta \Rightarrow \forall_{t\geq 0} : \|\mathbf{x}(t) - \mathbf{x}^*\| < \alpha \, \|\mathbf{x}(0) - \mathbf{x}^*\| \, e^{-\beta t}.$$

We are interested in a very specific type of equilibrium point, namely the *attractors* of the system. Attractors are equilibria toward which all neighboring trajectories converge, and their properties are described in Definition 1.2 [61].

Definition 1.2 (Attractor) A closed set $\mathcal{A}$ is an *attractor*, if it is the minimal set that satisfies the following two conditions:

1. $\mathcal{A}$ is an *invariant set*, that is,

$$\forall_{\mathbf{x}(0)\in\mathcal{A}} \, \forall_{t\geq 0} : \mathbf{x}(t) \in \mathcal{A}.$$

2. $\mathcal{A}$ attracts all trajectories that start sufficiently close to it, that is,

$$\exists_{\mathcal{U}\supset\mathcal{A}} \, \forall_{\mathbf{x}(0)\in\mathcal{U}} : \lim_{t\to\infty} \boldsymbol{x}(t) \in \mathcal{A}.$$

The largest such superset $\mathcal{U} \supset \mathcal{A}$ is called the *basin of attraction*.

Note that the definition of attractors only refers to closed sets, which can lead to complex structures, such as the well-known three-dimensional butterfly-shaped *Lorenz attractor* [42].

The Lorenz attractor is also called a strange attractor due to its unique behavior, which is governed by the equation system in Eq. (1.6). Sample trajectories in a three-dimensional phase space are illustrated in Fig. 1.9 for two different parameters $\rho = 14$ and $\rho = 28$.

$$\frac{dx}{dt} = \sigma \, (y - x) \qquad \frac{dy}{dt} = x \, (\rho - z) - y \qquad \frac{dz}{dt} = x \, y - \beta \, z. \qquad (1.6)$$

Figure 1.9 also illustrates the bifurcative influence of parameter ρ. While a small value of $\rho = 14$ in Fig. 1.9a results in a stable system that deterministically evolves toward a single equilibrium point, a large value of $\rho = 28$ in Fig. 1.9b causes

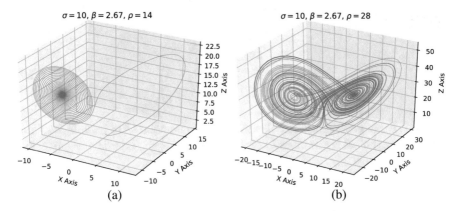

Fig. 1.9 Trajectory of the Lorenz attractor in three-dimensional phase space with $\sigma = 10$, $\beta = 8/3$: (**a**) stable orbit for $\rho = 14$, (**b**) chaotic behavior for $\rho = 28$

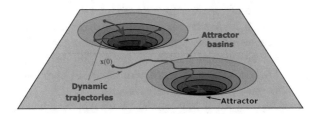

Fig. 1.10 Schematic illustration of attraction of trajectories in phase space

a highly complex and chaotic orbit, where the stable points repel the dynamic trajectory so that it never crosses itself. Note that although the system trajectory does not necessarily lead to a single point as an attractor, the entire structure itself forms a strange attractor in this specific case. In the following, however, we only consider point attractors, although several other types of attractors, such as cycle attractors and torus attractors, exist as well.

The general dynamic properties of attractors in the phase space are illustrated in Fig. 1.10. When, the initial state $\mathbf{x}(0)$ lies within the basin of an attractor, its trajectory will always lead it toward the attractor. However, if $\mathbf{x}(0)$ initially lies outside of the attractor's basin, the system state can only reach the attractor if its trajectory enters the attractor's basin. If the system always converges to certain attractors for any initial state $\mathbf{x}(0)$, these attractors are called global attractors; otherwise, as in the cases depicted in Fig. 1.10, these attractors are called local attractors.

To assess the stability of an attractor, its Jacobian matrix is computed, which contains the first-order partial derivatives of $\partial f_i(\mathbf{x})/\partial x_j$ over all $i, j = 1, \ldots, N$ as expressed in Eq. (1.7).

$$
J_{\mathbf{F(x)}} = \begin{pmatrix} \frac{\partial f_1(\mathbf{x})}{\partial x_1} & \cdots & \frac{\partial f_1(\mathbf{x})}{\partial x_N} \\ \vdots & & \vdots \\ \frac{\partial f_M(\mathbf{x})}{\partial x_1} & \cdots & \frac{\partial f_N(\mathbf{x})}{\partial x_N} \end{pmatrix}.
\tag{1.7}
$$

Then, it can be demonstrated that the equilibrium point $\mathbf{x}^\star$ is a *stable attractor*, if the real part of all eigenvalues λ_j of the matrix $J_{\mathbf{F(x)}}\big|_{\mathbf{x}=\mathbf{x}^\star}$ is strictly negative; that is, $\forall_j \operatorname{Re}\{\lambda_j\} < 0$.

1.4.2 Dynamic Systems Under the Influence of Noise

Thus far, we have discussed several fundamental definitions describing dynamic systems. In the following, we extend our formulation to include the concept of noise in the dynamics. Real-world data, regardless of whether they are obtained from biological or engineered systems, usually do not exhibit the "smooth" dynamics described in Sect. 1.4.1, but are rather faced with stochastic perturbations caused by transmission errors or noisy communication channels.

We now extend Eq. (1.4) to include a noise vector η in the dynamics of $\mathbf{x}$.

$$
\frac{d\mathbf{x}}{dt} = \mathbf{F}(\mathbf{x}) + \eta.
\tag{1.8}
$$

Here, η is an N-dimensional vector consisting of independent and identically distributed (i.i.d.) white noise terms η_i that are all zero-mean Gaussian random variables with standard deviation σ_i, $i = 1, \ldots, N$. Note that if the noise introduced by η is sufficiently small, the system still converges to an attractor in the same way as in Sect. 1.4.1, as long as the state vector remains within its basin of attraction despite the perturbations. Such types of stochastic differential equation systems can be solved numerically using the Euler–Maruyama method [22].

1.4.3 Relationship Between Fluctuation and Its Response

Before discussing the formulation of the Yuragi model, we introduce another related concept. For a classical thermodynamic system in physics, the fluctuation-dissipation theorem [33] describes the direct relationship between the response of a system when it is in equilibrium and when an external force is applied. However, the underlying assumption is that the metric of the system is a thermodynamic quantity. In [56], the authors revealed via biological experiments that the speed of fluorescence evolution of proteins in bacteria has a positive correlation with the phenotypic fluctuation of fluorescence in clone bacteria. They formulated a mathematical model of a biological system with a variable for a measurable quantity x that describes the system state and that can be influenced by parameter a.

Let $E[x_a]$ be the average of x under the influence of a. Then, if an external force is applied such that $a \rightarrow a + \Delta a$, the change in the average value of x is proportional to its variance at the initial parameter value a.

$$\frac{E[x_{a+\Delta a}] - E[x_a]}{\Delta a} \propto \text{Var}[x_a]. \tag{1.9}$$

The underlying assumption of this relationship is that x is approximately Gaussian distributed. The formulation of Eq. (1.9) is similar to the fluctuation-dissipation theorem; however, it is also valid for non-thermodynamic quantities and cases in which the fluctuation-dissipation theorem is not applicable at all. This fluctuation-response relationship can be considered a fundamental model that can help explain how the existence of noise enhances the adaptability of a dynamic system, as described in Sect. 1.5.

Based on this concept of fluctuation and its response, a model framework was proposed in [36] for selecting different networks and services based on the robustness of each network's performance. In this model, selection is performed based on observations of the system's responsiveness to inherent fluctuations. Due to the explicit utilization of the inherent fluctuations in the system, the proposed selection scheme can operate smoothly to select the most suitable and robust network.

1.5 Yuragi Model for Attractor Selection

In this section we introduce the Yuragi model, which is the main topic of the following chapters. The first model was introduced by Kashiwagi et al. in [27]. Before presenting the details of the model, we first consider its basic dynamic behavior as an adaptive feedback-based control mechanism that is driven by the interaction of two dynamic equations.

First, let us assume a system that must select among N possible choices. In this section, we only consider the principle mechanism with $N = 2$; however, in the following chapters, larger values of N are also considered in the discussion of realistic networking applications.

The basic mathematical model consists of two parts: the system state vector and activity. The state vector is expressed as a time-dependent vector $\mathbf{x}(t)$, and its dynamics is defined by a system of N stochastic differential equations that we express as $\mathbf{F}(\mathbf{x})$. The activity controls the level of trade-off between "exploration and exploitation" and is a function that maps the N-dimensional state vector to a non-negative real value (e.g., within the interval [0, 1]). Finally, $\boldsymbol{\eta}$ is a zero-mean N-dimensional Gaussian random variable. We now define the basic dynamics of the

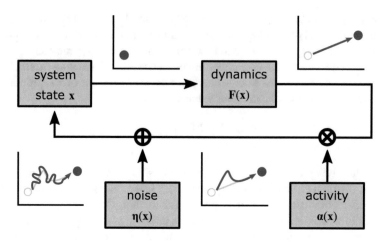

Fig. 1.11 Attractor selection illustrated as a noise-driven control loop influenced by activity

system state, as expressed in Eq. (1.10).

$$\frac{d\mathbf{x}}{dt} = \mathbf{F}(\mathbf{x})\,\alpha(\mathbf{x}) + \boldsymbol{\eta}. \tag{1.10}$$

Before providing a more detailed explanation of function $\mathbf{F}$ in Sects. 1.5.1–1.5.3, we first examine the fundamental behavior of the system in Eq. (1.10). It can be seen that there is permanent noise influence due to $\boldsymbol{\eta}$. Function $\mathbf{F}$ must be defined according to the objectives of the system, and characterizes the attractors to which the system state will converge, regardless of the small perturbations introduced by $\boldsymbol{\eta}$. However, as $\alpha(\mathbf{x})$ approaches 0, the influence of the first summand in Eq. (1.10) decreases, leaving the entire dynamics influenced by only the noise term, and a random walk in the phase space is performed. The basic behavior of attractor selection can thus be regarded as a noise-driven control loop as illustrated in Fig. 1.11.

In the following, we consider several examples of how this model has been defined in practical applications. Namely, we present three approaches to attractor selection that were proposed in [18, 19, 27] in various contexts. The commonality of these approaches is that they utilize the model described in Eq. (1.10). However, all three approaches define their attractors through individual functions $\mathbf{F}$ and activity α depending on their respective model scenarios.

1.5.1 Adaptive Response of Gene Network to Nutrient Availability

In [27], the attractor selection model was designed based on experiments of bistable switching behavior of two mutually inhibitory operons in *Escherichia coli* bacteria cells. An operon is a unit in the protein transcription process that creates messenger RNA (mRNA) and consists of an operator, promoter, and associated structural genes. In general, cells change their gene expression through their regulatory network, which occurs in response to environmental changes or external signals. The regulatory network is essentially a network of activating or inhibiting genes, and is controlled through specialized signal transduction networks to quickly react to frequently occurring signals. Thus, a specific reaction in the transduction network is activated if a certain event in the environment is detected. However, the number of all possible environmental changes is so large that a transduction pathway cannot exist for all cases. In [27], the authors investigated how adaptation toward a stable expression state is nevertheless achieved.

To study this phenomenon, Kashiwagi et al. [27] used a synthetic gene network composed of two mutually inhibitory operons that were attached with a green fluorescence protein and red fluorescence protein to visualize their dynamics. Exposure to a neutral medium resulted in a single monostable attractor, where the mRNA concentrations of both operons exhibited only weak expression. In contrast, when the network was exposed to a medium that was lacking either one of the key nutrients, it resulted in a bistable condition with two attractors: one with high green and low red fluorescence expression and one with high red and low green fluorescence expression. Thus, the former single stable attractor with weak expression became unstable, and fluctuations due to the noise inherent in gene expression caused the system to shift toward either one of the attractors due to their mutually inhibitory behavior.

A mathematical model of this phenomenon was derived in [27], describing the dynamics of the mRNA concentrations x_i of each operon $i = 1, 2$.

$$\frac{dx_1}{dt} = \frac{S(\alpha)}{1 + x_2^2} - D(\alpha) x_1 + \eta_1 \qquad \frac{dx_2}{dt} = \frac{S(\alpha)}{1 + x_1^2} - D(\alpha) x_2 + \eta_2. \qquad (1.11)$$

The functions $S(\alpha)$ and $D(\alpha)$ denote the rate coefficients of the synthesis and degradation of cell growth, respectively, and can be defined over a function α, which represents the cell activity.

$$S(\alpha) = \frac{6\alpha}{2 + \alpha} \qquad\qquad D(\alpha) = \alpha. \qquad (1.12)$$

The activity dynamics is explained in the supplemental material of [27] and defined using a classical model for the cell growth rate [53].

$$\frac{d\alpha}{dt} = \frac{P}{\prod\limits_{i=1}^{N} \left[\left(\frac{\theta_i}{x_i + N_i} \right)^{n_i} + 1 \right]} - C\alpha. \qquad (1.13)$$

An example of a numerical simulation is presented in Fig. 1.12 using the same parameters as in [27]. Here, the production and consumption rates of activity α are

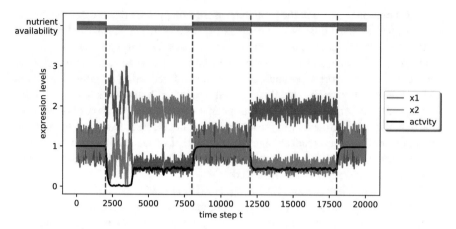

Fig. 1.12 Simulation of dynamics of two mutually inhibitory operons following the model in [27]

$P = C = 0.01$, respectively. $N_i \in \{0, 10\}$ represents the external supplementation of the respective nutrients, $\theta_i = 2$ are the thresholds of nutrient production, and $n_i = 5$ are their sensitivities for $i = 1, 2$. The bars at the top of the figure represent the nutrient availability over time, while the lower part of the figure depicts the expression levels of the two operons x_1 and x_2. We note that roughly between time steps 2000 and 4000, the system state lies outside the basin of the proper attractor. Thus, the activity decreases to 0 (black line), and a random walk is performed until the system finally converges to the attractor, causing activity to increase to about 0.5.

This model has been extended to multiple dimensions and applied to selection in various networking problems [37–40], as discussed in greater detail in Chap. 3.

1.5.2 Modeling the Interactions of Gene Expression and Metabolic Flux

An alternative model for the dynamics of a gene regulatory network was proposed by Furusawa and Kaneko in [19]. In this model, the dynamics of the protein concentrations x_i are described by the differential equation system in Eq. (1.14). However, in this model, the growth rate of the cell α (written as $v_g(t)$ in [19]) is obtained from the metabolic flux in the cell caused by the current expression level in the on/off type of gene regulation.

$$\frac{dx_i}{dt} = \left[f\left(\sum_{j=1}^{M} w_{ij} x_j - \theta \right) - x_i \right] \alpha(\mathbf{x}, \mathbf{y}) + \eta_i \quad i = 1, \ldots, M. \quad (1.14)$$

The weights are discrete values $w_{ij} \in \{-1, 0, 1\}$ depending on whether protein i inhibits, is neutral to, or activates protein j, and the activation is performed by a sigmoid function f using a bias term θ. The focus in [19] was on the investigation of the cell growth dynamics. In their experiments, Furusawa and Kaneko used $M = 96$ and randomly selected weights w_{ij} with positive auto-regulatory values of $w_{ii} = 1$. Their results yielded an optimal range for the noise amplitude to achieve high cell growth rates without converging to suboptimal attractors. Cell growth is influenced by the dynamics of the concentration of the ith substrate in the metabolic network:

$$\frac{dy_i}{dt} = \varepsilon \sum_{j=1}^{M} \sum_{k=1}^{N} Con(k, j, i) \, x_j \, y_k$$

$$i = 1, \ldots, N \qquad (1.15)$$

$$- \varepsilon \sum_{j=1}^{M} \sum_{k=1}^{N} Con(i, j, k) \, x_j \, y_i + D \, (Y_i - y_i),$$

where $Con(i, j, k)$ is a binary variable representing the metabolic reaction, where the ith substrate to the kth substrate is catalyzed by the jth protein. Term D is the diffusion coefficient, and Y_i is the extracellular concentration of the ith substrate. Cell growth itself is assumed to be such that it requires at least r metabolites for growth and is selected as

$$\alpha \propto \min \{y_1, \ldots, y_r\}.$$

Note that in [19], the weights remain constant as they represent the evolved signal transduction pathway for expressing certain genes. A study of the robustness to noise in the gene expression of this model is provided in [26], where mutations in the pathways reflect evolutionary changes. For the sake of simplicity, we assume a generalized model inspired by Kaneko [26], as given in Eq. (1.16).

$$\frac{dx_i}{dt} = \left[\tanh \left(\beta \sum_{j=1}^{M} w_{ij} x_j \right) - x_i \right] \alpha(\mathbf{x}, \mathbf{y}) + \eta_i \quad i = 1, \ldots, M. \qquad (1.16)$$

Unlike in the model described in Sect. 1.5.1, the activity is dependent on $\mathbf{x}$ and $\mathbf{y}$. The constant β represents the sensitivity of activation. In [26], the weights were adapted over several generations taking into account the fitness of the system. Similar equations to those in Eq. (1.16) are also often used as models of artificial neurons in recurrent neural networks [57], where the weights are adapted during the learning iterations and can take continuous values. Additional details about this model and its applications to virtual network control are provided in Chap. 4.

1.5.3   Gaussian Mixture Model Attractors

When the possible target values are known in advance, the attractors can be designed in an appropriate way. Such an approach was proposed in [18] to control a two-link robotic arm. A Gaussian mixture model was used to formulate the global potential function $F(\mathbf{x})$, which consists of the superposition of N-dimensional Gaussian functions around K predetermined target candidate values $\boldsymbol{\xi}_1, \ldots, \boldsymbol{\xi}_K$,

$$F(\mathbf{x}) = -\frac{1}{K v \sqrt{2\pi}} \sum_{i=1}^{K} e^{-\frac{1}{2v^2} \|\boldsymbol{\xi}_i - \mathbf{x}\|^2} \qquad (1.17)$$

and the dynamics is obtained as derivative $\partial F(\mathbf{x})/\partial x_i$ for $i = 1, \ldots, N$.

Thus, the target vectors $\boldsymbol{\xi}_i$ form attractors with standard deviation v, which influences the attractor basins, and the activity can be formulated as inversely proportional to v, which makes the potential function steeper when activity is high and flatter when it is low. The authors of [18] proposed further extensions to rearranging the attractors using a clustering technique, where high activity can be maintained. Unlike in the methods described in Sects. 1.5.1 and 1.5.2, activity is updated by the sum of an instantaneous achievement of the robot with a decaying moving average of the previous activity values.

1.6   Conclusion

In this chapter, we examined how noise is utilized in biological systems to adapt to changes in the environment. We observed that many biological systems often use simple control loop mechanisms that can operate robustly when failures occur, as they are not bound to individual control units, but rather, each entity in the system contributes to finding a solution.

Specifically, we introduced Yuragi control, which uses a mathematical formulation of system dynamics using attractors. Depending on the feedback it receives, the Yuragi controller can control the balance between exploration and exploitation of the current solution. In the following chapters, additional information is presented to explain how the Yuragi method can be implemented in the control and traffic engineering of information networks. The main benefit we expect to observe by applying the attractor selection concept to information networks is not a system that can perform optimally in a controlled environment, but a system that can operate without much configuration in a robust and self-adaptive way in an unknown, unpredictable, and fluctuating environment.

In this chapter, we presented several variations of implementing Yuragi control through the attractor selection. When transferring attractor selection from biology to information networks, we must carefully consider how to map the corresponding components and perform modifications to make its application feasible and meaningful.

The first point to be addressed is the appropriate definition of the activity or cell growth. In the biological models discussed thus far, there is a functional relationship between the environmental input, current system state, and activity. However, this is because the models focus on single cells only and the activity or growth rate can be mapped directly to the cell's gene expression or metabolic flow. When applying the attractor selection method to an information network setting, we must be aware of what the cell, and ultimately the state vector $\mathbf{x}$ correspond to. If we consider that a state consists of certain parameter settings on an end-to-end connection that is oblivious to the state of the entire network, our decisions may have implications that are unknown a priori, and the activity and influence of the current selection of attractors may be only quantified by means of feedback. In fact, the growth rate can be defined as a quantity perceived from the end user's viewpoint, consisting of either objective Quality of Service (QoS) or subjective Quality of Experience (QoE) metrics.

Another important problem is the interpretation of the noise terms. In biological systems, noise is an inherent feature, and its amplitude may be considered independent of the current state of gene expression or activity. This is reflected in using a constant and independent noise term η; however, in designing an engineered system, we take into account the following two possibilities for including fluctuations: (a) noise is system-inherent and can only be observed, but not influenced, and (b) the application can modify the noise amplitude to increase the robustness and speed of adaptation. At this point, we only focus on (a) and assume that noise is not influenced by our approach. Attractors are by definition representations of the stable states of an entire system. In neural networks, attractors correspond to the patterns that are stored within the network for classification, such as the minima of the energy function in *Hopfield networks* [24].

The future of networking must accommodate a growing number of services and users that exhibit *self-star* capabilities [4]. Many of the bio-inspired approaches discussed in this chapter only focus on individual specific mechanisms to solve a particular problem. However, from the future Internet perspective, we also need to take a step back to see how these existing bio-inspired solutions can be integrated to meet the many requirements of the future Internet [5].

References

1. Aarts, E., Korst, J.: Simulated Annealing and Boltzmann Machines. Wiley, Chichester (1989)
2. Ab Wahab, M.N., Nefti-Meziani, S., Atyabi, A.: A comprehensive review of swarm optimization algorithms. PLoS ONE **10**(5), 1–36 (2015)
3. Aono, M., Hara, M., Aihara, K.: Amoeba-based neurocomputing with chaotic dynamics. Commun. ACM **50**(9), 69–72 (2007)
4. Babaoglu, O., Jelasity, M., Montresor, A., Fetzer, C., Leonardi, S., van Moorsel, A., van Steen, M.: The self-star vision. In: Self-star Properties in Complex Information Systems, pp. 1–20. Springer, Berlin (2005)

5. Balasubramaniam, S., Leibnitz, K., Lio, P., Botvich, D., Murata, M.: Biological principles for future internet architecture design. IEEE Commun. Mag. **49**(7), 44–52 (2011)
6. Birn, R.M.: The role of physiological noise in resting-state functional connectivity. NeuroImage **62**(2), 864–870 (2012)
7. Blake, R., Logothetis, N.K.: Visual competition. Nat. Rev. Neurosci. **3**(1), 13–21 (2002)
8. Bonabeau, E., Dorigo, M., Theraulaz, G.: Swarm Intelligence: From Nature to Artificial Systems. Oxford University Press, Oxford (1999)
9. Brin, S., Page, L.: The anatomy of a large-scale hypertextual web search engine. Comput. Netw. ISDN Syst. **30**(1), 107–117 (1998)
10. Burda, Z., Duda, J., Luck, J.M., Waclaw, B.: Localization of the maximal entropy random walk. Phys. Rev. Lett. **102**, 160602 (2009)
11. Deco, G., Romo, R.: The role of fluctuations in perception. Trends Neurosci. **31**(11), 591–598 (2008)
12. Di Caro, G., Dorigo, M.: The Ant Colony Optimization Meta-Heuristic, pp. 250–285. McGraw-Hill, London (1999)
13. Di Caro, G., Ducatelle, F., Gambardella, L.M.: Anthocnet: an adaptive nature-inspired algorithm for routing in mobile ad hoc networks. Eur. Trans. Telecommun. **16**(5), 443–455 (2005)
14. Dorigo, M., Stützle, T.: Ant Colony Optimization. A Bradford Book. The MIT Press, Cambridge (2004)
15. Dressler, F.: Self-Organization in Sensor and Actor Networks. Wiley, New York (2007)
16. Ermentrout, G.B., Galán, R.F., Urban, N.N.: Reliability, synchrony and noise. Trends Neurosci. **31**(8), 428–434 (2008)
17. Farooq, M., Di Caro, G.A.: Routing Protocols for Next-Generation Networks Inspired by Collective Behaviors of Insect Societies: An Overview, pp. 101–160. Springer, Berlin (2008)
18. Fukuyori, I., Nakamura, Y., Matsumoto, Y., Ishiguro, H.: Flexible control mechanism for multi-DOF robotic arm based on biological fluctuation. In: 10th International Conference on the Simulation of Adaptive Behavior (SAB'08), Osaka (2008)
19. Furusawa, C., Kaneko, K.: A generic mechanism for adaptive growth rate regulation. PLoS Comput. Biol. **4**(1), e3 (2008)
20. Gammaitoni, L., Hänggi, P., Jung, P., Marchesoni, F.: Stochastic resonance. Rev. Mod. Phys. **70**, 223–287 (1998)
21. González, M.C., Hidalgo, C.A., Barabási, A.L.: Understanding individual human mobility patterns. Nature **453**(7196), 779–782 (2008)
22. Higham, D.J.: An algorithmic introduction to numerical simulation of stochastic differential equations. SIAM Rev. **43**(3), 525–546 (2001)
23. Holland, J.: Adaptation in Natural and Artificial Systems. MIT Press, Cambridge (1992)
24. Hopfield, J.: Neural networks and physical systems with emergent collective computational abilities. Proc. Natl. Acad. Sci. USA **79**(8), 2554–2558 (1982)
25. Kaneko, K.: Life: An Introduction to Complex Systems Biology. Springer, Berlin (2006)
26. Kaneko, K.: Evolution of robustness to noise and mutation in gene expression dynamics. PLoS ONE **2**(5), e434 (2007)
27. Kashiwagi, A., Urabe, I., Kaneko, K., Yomo, T.: Adaptive response of a gene network to environmental changes by fitness-induced attractor selection. PLoS ONE **1**(1), e49 (2006)
28. Kashtan, N., Alon, U.: Spontaneous evolution of modularity and network motifs. Proc. Natl. Acad. Sci. **102**(39), 13773–13778 (2005)
29. Kashtan, N., Noor, E., Alon, U.: Varying environments can speed up evolution. Proc. Natl. Acad. Sci. **104**(34), 13711–13716 (2007)
30. Kim, S.J., Aono, M., Hara, M.: Tug-of-war model for multi-armed bandit problem. In: Calude, C.S., Hagiya, M., Morita, K., Rozenberg, G., Timmis, J. (eds.) Unconventional Computation, pp. 69–80. Springer, Berlin (2010)
31. Kim, S.J., Naruse, M., Aono, M., Ohtsu, M., Hara, M.: Decision maker based on nanoscale photo-excitation transfer. Sci. Rep. **3**(1), 2370 (2013)

32. Kish, L., Granqvist, C.: Noise in nanotechnology. Microelectron. Reliab. **40**(11), 1833–1837 (2000)
33. Kubo, R.: The fluctuation-dissipation theorem. Rep. Prog. Phys. **29**(1), 255–284 (1966)
34. Kuroda, K., Kato, H., Kim, S.J., Naruse, M., Hasegawa, M.: Improving throughput using multi-armed bandit algorithm for wireless LANs. Nonlinear Theory Appl. IEICE **9**(1), 74–81 (2018)
35. Leibnitz, K., Hoßfeld, T., Wakamiya, N., Murata, M.: Peer-to-peer vs. client/server: Reliability and efficiency of a content distribution service. In: 20th International Teletraffic Congress (ITC-20), Ottawa, pp. 1161–1172 (2007)
36. Leibnitz, K., Murata, M.: Attractor selection and perturbation for robust networks in fluctuating environments. IEEE Netw. **24**(3), 14–18 (2010)
37. Leibnitz, K., Wakamiya, N., Murata, M.: Biologically inspired self-adaptive multi-path routing in overlay networks. Commun. ACM **49**(3), 62–67 (2006)
38. Leibnitz, K., Wakamiya, N., Murata, M.: Resilient multi-path routing based on a biological attractor selection scheme. In: 2nd International Workshop on Biologically Inspired Approaches to Advanced Information Technology (BioAdit'06). Springer, Osaka (2006)
39. Leibnitz, K., Wakamiya, N., Murata, M.: Self-adaptive ad-hoc/sensor network routing with attractor-selection. In: IEEE GLOBECOM. IEEE, San Francisco (2006)
40. Leibnitz, K., Wakamiya, N., Murata, M.: A bio-inspired robust routing protocol for mobile ad hoc networks. In: 16th International Conference on Computer Communications and Networks (ICCCN'07), Honolulu, pp. 321–326 (2007)
41. Leibnitz, K., Yomo, T., Murata, M.: Attractor selection as self-adaptive control mechanism for communication networks. In: Xiao, Y. (ed.) Bio-Inspired Computing and Networking, chap. 14, pp. 369–389. CRC Press, Boca Raton (2011)
42. Lorenz, E.N.: Deterministic nonperiodic flow. J. Atmos. Sci. **20**, 130–141 (1963)
43. Ma, J., Hasegawa, S., Kim, S.J., Hasegawa, M.: A reinforcement-learning-based distributed resource selection algorithm for massive IoT. Appl. Sci. **9**(18), 3730 (2019)
44. Masuda, N., Porter, M.A., Lambiotte, R.: Random walks and diffusion on networks. Phys. Rep. **716–717**, 1–58 (2017)
45. McDonnell, M., Ward, L.: The benefits of noise in neural systems: bridging theory and experiment. Nat. Rev. Neurosci. **12**, 415–425 (2011)
46. McDonnell, M., Stocks, N., Pearce, C., Abbott, D.: Optimal information transmission in nonlinear arrays through suprathreshold stochastic resonance. Phys. Lett. A **352**, 183–189 (2006)
47. McDonnell, M., Stocks, N., Pearce, C., Abbott, D.: Stochastic Resonance. Cambridge University Press, Cambridge (2008)
48. Mitaim, S., Kosko, B.: Adaptive stochastic resonance. Proc. IEEE **86**(11), 2152–2183 (1998)
49. Mizutani, S., Arai, K., Davis, P., Wakamiya, N., Murata, M.: Noise-assisted distributed detection in sensor networks. AIP Conf. Proc. **922**(1), 611–614 (2007)
50. Murata, T., Hamada, T., Shimokawa, T., Tanifuji, M., Yanagida, T.: Stochastic process underlying emergent recognition of visual objects hidden in degraded images. PLoS ONE **9**(12), 1–32 (2014)
51. Nakagaki, T., Yamada, H., Tóth, Á.: Maze-solving by an amoeboid organism. Nature **407**(470) (2000)
52. Naruse, M., Berthel, M., Drezet, A., Huant, S., Aono, M., Hori, H., Kim, S.J.: Single-photon decision maker. Sci. Rep. **5**(1), 13253 (2015)
53. Nielsen, J., Villadsen, J.: Bioreaction Engineering Principles. Plenum Press, New York (1994)
54. Otokura, M., Leibnitz, K., Shimokawa, T., Murata, M.: Evolutionary core-periphery structure and its application to network function virtualization. Nonlinear Theory Appl. IEICE **7**(2), 202–216 (2016)
55. Otokura, M., Leibnitz, K., Koizumi, Y., Kominami, D., Shimokawa, T., Murata, M.: Evolvable virtual network function placement method: mechanism and performance evaluation. IEEE Trans. Netw. Serv. Manag. **16**(1), 27–40 (2019)
56. Sato, K., Ito, Y., Yomo, T., Kaneko, K.: On the relation between fluctuation and response in biological systems. Proc. Natl. Acad. Sci. USA **100**(24), 14086–14090 (2003)

57. Schmidhuber, J.: Deep learning in neural networks: an overview. Neural Netw. **61**, 85–117 (2015)
58. Sims, D.W., Southall, E.J., Humphries, N.E., Hays, G.C., Bradshaw, C.J.A., Pitchford, J.W., James, A., Ahmed, M.Z., Brierley, A.S., Hindell, M.A., Morritt, D., Musyl, M.K., Righton, D., Shepard, E.L.C., Wearmouth, V.J., Wilson, R.P., Witt, M.J., Metcalfe, J.D.: Scaling laws of marine predator search behaviour. Nature **451**(7182), 1098–1102 (2008)
59. Sinatra, R., Gómez-Gardeñes, J., Lambiotte, R., Nicosia, V., Latora, V.: Maximal-entropy random walks in complex networks with limited information. Phys. Rev. E **83**, 030103 (2011)
60. Stocks, N.G.: Suprathreshold stochastic resonance in multilevel threshold systems. Phys. Rev. Lett. **84**, 2310–2313 (2000)
61. Strogatz, S.H.: Nonlinear Dynamics and Chaos. Westview Press, Cambridge (1994)
62. Viterbi, A.J.: CDMA: Principles of Spread Spectrum Communication. Prentice Hall, Upper Saddle River (1995)
63. Wakamiya, N., Leibnitz, K., Murata, M.: Noise-assisted control in information networks. In: 2007 Frontiers in the Convergence of Bioscience and Information Technologies, pp. 833–838 (2007)
64. Wedde, H.F., Farooq, M., Zhang, Y.: Beehive: An efficient fault-tolerant routing algorithm inspired by honey bee behavior. In: Dorigo, M., Birattari, M., Blum, C., Gambardella, L.M., Mondada, F., Stützle, T. (eds.) Ant Colony Optimization and Swarm Intelligence, pp. 83–94. Springer, Berlin (2004)
65. Weinstein, S., Pavlic, T.P.: Noise and Function, pp. 174–198. Cambridge University Press, Cambridge (2017)
66. White, J.A., Rubinstein, J.T., Kay, A.R.: Channel noise in neurons. Trends Neurosci. **23**(3), 131–137 (2000)
67. Wiesenfeld, K., Jaramillo, F.: Minireview of stochastic resonance. Chaos Interdiscip. J. Nonlinear Sci. **8**(3), 539–548 (1998)
68. Yanagida, T., Ueda, M., Murata, T., Esaki, S., Ishii, Y.: Brownian motion, fluctuation and life. Biosystems **88**(3), 228–242 (2007)

Chapter 2
Functional Roles of Yuragi in Biosystems

Toshio Yanagida and Tsutomu Murata

Abstract What are the underlying principles that explain how complex biosystems work in such a remarkably energy-saving and flexible manner? In this chapter we explore this question based on our research of muscle and brain, both of which are typically complex biosystems. Our state-of-the-art imaging technology of direct observation of individual molecular motor motion has revealed the surprising fact that muscle contraction is produced by utilizing the Brownian motion of molecular motors in a very skillful manner. This indicates that the muscle can utilize thermal fluctuations of molecules in an effective way, such that the energy efficiency of muscle is extremely high compared to artificial energy conversion systems. Although the research methods are quite different, we have also observed analogous findings in human brain function. Our research on human visual recognition showed that the time taken for recognizing a difficult figure follows the same exponential function as the "Arrhenius equation," which describes how the rate of a chemical reaction is driven by thermal fluctuation. This finding, as well as our related modeling, strongly supports the idea that stochastic activity, possibly resting-level spontaneous activity, may also help make human recognition flexible and energy saving. Based on these findings, we would argue that fluctuations in biosystems, both thermal fluctuation of motor molecules in muscle and stochastic activity of neurons in the brain, play an essential role in flexible and energy-saving functioning. To emphasize these positive aspects of fluctuations in biosystems, we would propose the concept of "Yuragi," which is a word of Japanese origin with the meaning of fluctuations for flexible adaptation to the environment. We would suggest that the utilization of Yuragi is one of the principles for efficiency and flexibility of biosystems.

T. Yanagida (✉) · T. Murata
Center for Information and Neural Networks, National Institute of Information and
Communications Technology, Suita, Osaka, Japan
e-mail: yanagida@nict.go.jp; benmura@nict.go.jp

© Springer Nature Singapore Pte Ltd. 2021
M. Murata, K. Leibnitz (eds.), *Fluctuation-Induced Network Control and Learning*,
https://doi.org/10.1007/978-981-33-4976-6_2

2.1   Introduction

Artificial intelligence (AI) recently defeated the Go game world champion Lee Sedol and became a big topic. The progress of AI technology in recent years is striking. However, AI is still far from reaching the capability of the human brain. One problem is energy cost. In 2016, DeepMind's αGO consumed an estimated 200,000 W when it played against Lee Sedol [26]. On the other hand, how much energy did Lee Sedol's brain consume? We have developed MRI (magnetic resonance imaging) and MRS (magnetic resonance spectroscopy) technologies that measure brain temperature with an accuracy of 0.1°. Then, the temperature change of the brain was measured when the brain was thinking deeply and when not thinking about anything. And from this temperature change, we estimated the energy consumption when the brain was actively thinking. Surprisingly, the result was that the brain additionally consumed only 1 W during activity (up to 20 W even including the energy consumption in the cell metabolism and default mode fluctuations in the resting state) [26], see Fig. 2.1. It is 1/200,000 compared to AI. In 2017, αGO was specialized in the Go game to significantly reduce the energy costs (αGO Zero), but still a few kilowatts. The human brain does not specialize in specific tasks such as Go, but works for various tasks. Still, the energy cost is 1 W. How can the brain work with so little energy? The brain is by no means a simple organization. In the cerebrum, 14 billion nerve cells (neurons) form a complex network. One neuron is connected to other neurons by thousands of synapses. There are tens of trillions of connections in the entire brain. The number of these combinations is 2 to the power of 10^{13}, even if it is simply turned on and off. If this enormous combination is to be calculated strictly with supercomputers, a huge amount of energy would be required, probably the energy of more than hundreds of millions of nuclear power plants to operate many supercomputers. But the brain works at only 1 W. Solving this energy-saving mechanism of the brain is expected to make a major contribution

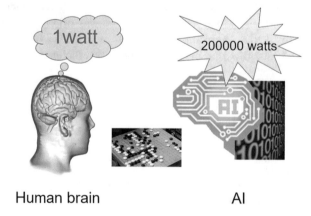

Fig. 2.1 Energy consumption of human brain and AI

to the elucidation of the mechanism of life, and at the same time, it will greatly contribute to IT development such as AI that will continue to develop and become more complex.

This chapter introduces our research, focusing on the fundamental differences between artificial and biological machines. We have been paying attention to fluctuations (noise). Fluctuation has a negative image from the viewpoint of artificial machines, but here we will show that it works positively in biosystems, taking biological molecular machines and human brain as examples. So, instead of an English word, fluctuation, we will use a Japanese word, "Yuragi," which has a positive nuance that represents more than just fluctuation.

2.2 How Muscle Works

2.2.1 Biological Molecular Motor in Muscle

Our body is made up of cells. Many molecular machines made of protein molecules are working in the cells. They play various roles essential for cellular activities in reading genetic information of DNA, synthesizing proteins, sending signals to cells by receiving stimuli from the outside, digesting food and converting it into chemical energy of ATP, and so on. And the molecular machine responsible for the movement, which is the symbol of living things, is molecular motors. Here, we will introduce our research on a molecular motor that works in muscle. Figure 2.2 shows the structure of the striated muscle. Muscle contraction occurs due to the interaction of actin with the globular domain of myosin molecule (hereinafter called myosin head). The energy for contraction is the chemical energy generated when ATP (adenosine triphosphate) is hydrolyzed into ADP (adenosine diphosphate) and Pi (phosphate). The problem is how the myosin head and actin convert chemical energy into mechanical energy to cause the flexible and energy-saving muscle contraction [9]. The problem has been studied extensively over the years, but no clear model based on direct experimental evidences has been proposed. This is because the movement of myosin head on actin filament has been indirectly inferred by observing from outside of muscle cells. Therefore, we have developed techniques for directly observing them and challenged this problem.

2.2.2 Single Molecule Imaging and Nano-Detection

We have developed a single molecule imaging and nano-detection technology that directly observes the movement of one myosin head and the hydrolysis reaction of ATP. Since the size of the myosin head is only 10–15 nm (nm $= 10^{-9}$ m), it cannot be observed with an optical microscope due to light diffraction in principle.

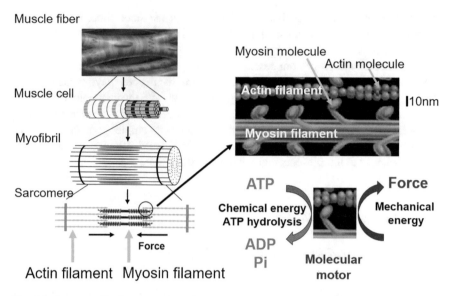

Fig. 2.2 Structure of striated muscle and myosin molecular motor. Muscle contraction is generated by myosin motors that convert the chemical energy of ATP hydrolysis into the mechanical energy with actin

The reason for using an optical microscope is to directly observe the movement of myosin head in an aqueous solution. We can observe the stars in the night sky if they shine no matter how many light years away. The myosin head should look like a star if it glows by labeling it with a fluorescent molecule. However, the fluorescence intensity emitted by a single dye molecule is extremely weak. The sensitivity of an ultra-sensitive camera is high enough to detect it but the problem is how to reduce background noise to create deep darkness. Dusts, stray light from lenses and filters, Raman scattering from water molecules, and so on generate huge background noise against the fluorescence from a single dye molecule. We used a local illumination method that illuminates only the fluorescent molecule and its immediate vicinity using a special light called evanescent wave generated by total internal reflection. We also removed dusts from the aqueous solution and carefully selected lenses and filters. And finally, we were able to reduce the background noise 2000-fold compared with conventional fluorescence microscopy. In 1995, we succeeded in directly observing single fluorescently labeled myosin heads and the ATP hydrolysis reaction by a single myosin head [5]. Furthermore, we have combined the nanometer detection technique that allows us to measure the movement and force of myosin heads with the accuracy of 1 nm and piconewton, respectively, [3, 10] with the single molecule imaging. Thus, we have succeeded in simultaneous observation of the ATP hydrolysis reaction and movement of myosin head (Fig. 2.3) [11].

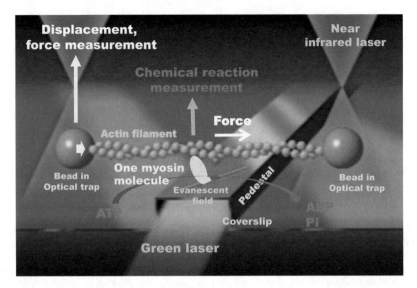

Fig. 2.3 Single molecule imaging and nanometer detection. The chemical reaction (ATP hydrolysis) and mechanical reaction (movement) of a single myosin motor are simultaneously observed

2.2.3   Bias Brownian Motion Model (Yuragi Model)

The single molecule imaging and nano-detection of the movement of myosin heads have led to very exciting results. We have found that during one ATP hydrolysis reaction, the myosin head fluctuates back and forth on the actin filament in which actin molecules are arranged at intervals of 5.5 nm and proceeds forward [4, 14]. The observed 5.5 nm steps should be caused by Brownian motion because they are not accompanied by the ATP hydrolysis. That is, chemical energy is not used for the 5.5 nm steps meaning that the muscle contraction is caused by thermal noise! Unlike macroscopic artificial machines, the myosin heads of tens of nanometers undergo Brownian motion in aqueous solution due to collisions of water molecules, so it is not surprising that this happens. But does the Brownian motion caused by the collision of water molecules have enough power to cause muscle contraction? Indeed, the average Brownian motion power is about a tenth smaller than the power required for muscle contraction. However, the power of Brownian motion is not constant and fluctuates greatly and so it is possible to use the Brownian motion that happens to fluctuate with large power. Even if Brownian motion provides sufficient power, another problem is that it is random and in order to use it for muscle contraction, it is necessary to bias the random motion in one direction. Recently the myosin head has been found to have a strain sensor that senses forward and backward Brownian motion [12]. This suggests that the myosin head moves rectifying random Brownian motion to the forward motion with the strain sensor. If this reaction takes place without energy, the muscle becomes a perpetual motion

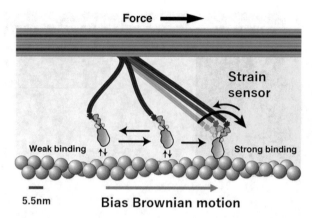

Fig. 2.4 Biased Brownian motion of myosin motor. The myosin motor fluctuates back and forth while weakly interacting with an actin filament by Brownian motion. When it happens to move forward, it binds strongly to actin by the action of strain sensor, which senses the direction of movement of myosin motor, and stopped the Brownian motion. In this way, random Brownian motion of myosin motor is biased in the forward direction and the directional motion is generated

machine. The chemical energy of ATP is not used to drive the movement of myosin, but it is used to operate the sensor, that is, to rectify the random Brownian motions to the forward movement, not inconsistent with the 2nd law of thermodynamics. We call this model Bias Brownian motion model, or Yuragi model (Fig. 2.4).

This model can be expressed by the Langevin equation (refer to Chap. 1) representing the Brownian motion with a bias term B added.

$$\frac{dx}{dt} = -\frac{1}{\rho}\frac{\partial U(x,t)}{\partial x}B + \sqrt{\frac{2kT}{\rho}}\,\eta(t).$$

We call this the Yuragi equation of molecular motor. Using computer simulation based on this equation, we have investigated whether the Yuragi model of molecular motor can improve our understanding of the basic mechanical properties of muscle [17, 18]. Many studies have been conducted on the molecular mechanism of muscles, but there has been no model that fully explains the mechanical characteristics of muscle in steady state and transient states based on direct experimental evidences. This "bottom-up" approach is becoming more and more common in muscle modeling literature, following the improvements in single molecule experimental techniques that allow for quantitative estimation of molecular properties, as described above. These "bottom-up" models are opposite to "top-down" models, which start from the muscle behavior to infer molecular properties. Bottom-up models instead try to conciliate the experimental evidences at the molecular level with these at the fiber or whole muscle level and can exploit the importance of thermal fluctuations in muscle contraction. We have found that a model [17] based on the Brownian motion of myosin heads along the actin filament (B term in the above equation), and where the ATP is not used to directly produce force in the actomyosin complex, can successfully explain the relationships between muscle contraction speed and force, and between efficiency and power. Furthermore, the model is able to simulate the responses of the muscle when mechanical perturbations are applied. We have also shown the importance of unbiased Brownian motion in the efficiency of muscle contraction [18], introducing it in an otherwise classic model (i.e., where the ATP energy is mainly used to generate force within the actomyosin complex). In fact, also in the bottom-up models, the classic approach is to define the actomyosin potential energy as a series of infinitely-steep minima. In this approach, the effect of Brownian motion is hidden in the ad-hoc definition of the rate constants for the transitions between stable states, used to simulate a desired behavior. Instead, we allow some degree of fluctuations even within the actomyosin stable states, which mathematically corresponds to define its potential energy with wider minima. The model shows that, including thermal fluctuations also in the stable states, the total efficiency of the contraction is increased, as well as the force per motor predicted by the model. The Yuragi model is a hopeful molecular mechanism that can solve the long-standing problem of how muscle works (Fig. 2.5). From the perspective of engineering, muscles are an ideal actuator that is highly efficient, energy saving, and responds flexibly to changes in the external world. These ideal properties of muscle as an actuator are realized by the positive effects of Yuragi as explained above. Recently there have been many reports that the mechanism that uses Yuragi works in various biological molecular machines that function in other type of molecular motors [2, 13, 20–22], Ca ion pump [24], gene expression

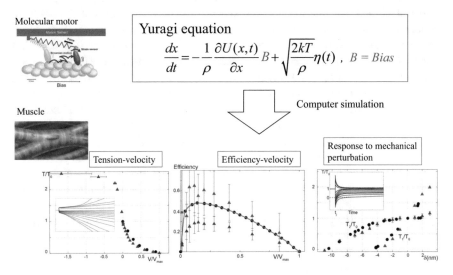

Fig. 2.5 Simulation of mechanical characteristics of muscle contraction based on the Yuragi equation. The steady state and transient reactions of muscle contraction are well explained by the bias Brownian motion

[1, 8], cell signal transduction [6, 7], etc. The Yuragi mechanism is now the common working principle of biological molecular machines.

2.3   How the Human Brain Recognizes Puzzling Figures by Means of Yuragi Activity

2.3.1   Yuragi Activity in the Human Brain

An outstanding enigma of the human brain is its low energy consumption in contrast to its high-level flexibility of information processing. The brain's superiority of energy efficiency over the contemporary computers is evident typically in "mind-sport" games such as Go and Shogi as introduced above in this chapter. In these games, solutions must be selected from an immense number of possibilities, which cannot be counted up in seconds even from the beginning of the universe. The computers must encode and process each possibility using an extremely small but certain amount energy so that a vast number of possibilities inevitably require huge energy in total (e.g., 200 kW). By contrast, energy requirements of the brain of a mind-sport player grow only a little, possibly by one more piece of chocolate cake. This evident difference in the energetics of computation demands us to reconsider the way how the brain processes information.

In this chapter we are proposing that Yuragi is a key concept to answer this problem. As Yuragi we refer to disordered neural activity that plays an important functional role in encoding and processing information of the brain. In neuroscience it is well known that a substantial proportion of neurons of the brain are intermittently active even without any particular stimulation [15, 16]. This kind of neural activity is spontaneous, disordered (almost stochastic), low firing-rate, and consuming only resting-level energy. In our proposition, the resting-level disordered neural activity provides the basis of Yuragi activity of the brain because they can keep potentially a vast number of functions through intermittent interactions among any possible functional ensembles although their activity magnitudes are under certain thresholds of exerting particular functions. Once a particular external stimulation is given, the input-driven activity instantly interacts with the preexisting Yuragi activity, which spontaneously provides various activity patterns one after another from a vast number of possibilities, to search for a solution candidate. Because this search process is caused by resting-level spontaneous activity, the energy consumption grows only a little from the resting state. There could also be an advantage for disorder of Yuragi activity to search for unexpected solutions making use of its actions free from expectation.

2.3.2   Psychophysical Experiment of Hidden-Figure Recognition

As a successful example of Yuragi concept application, here we show our experimental study which found stochastic property of the visual system in resolving difficult (puzzling) recognition tasks, and in the next subchapter we show a Yuragi model of this phenomenon. When we see a severely degraded image such as Fig. 2.6 for the first time, only meaningless black-white patterns are seen. After observing for a while, however, you suddenly recognize some familiar objects hidden in the image in an emergent manner even without being given any prior knowledge of the objects. (The not-degraded image as the answer is shown below in this chapter.) To recognize such hidden figures is challenging because these kinds of images do not keep most of information of the depicted objects owing to binarization, so that missing information of objects must be complemented for recognition. In general this complementation will cause an explosion of the number of possibility because many potential image features can be candidates to be complemented before the hidden objects are recognized. Calculations of the explosive possibility are energy consuming for the contemporary computer systems, while the brain can often solve such problems under normal observation in seconds or minutes as you might experience above.

To investigate the way how the human visual system recognizes hidden figures, we conducted a psychophysical experiment in which we measured the time needed to recognize correct objects hidden in binarized images [19]. To avoid any prior knowledge and categorical inference of the hidden objects, we created 90 binarized

Fig. 2.6 An example of hidden-figure image

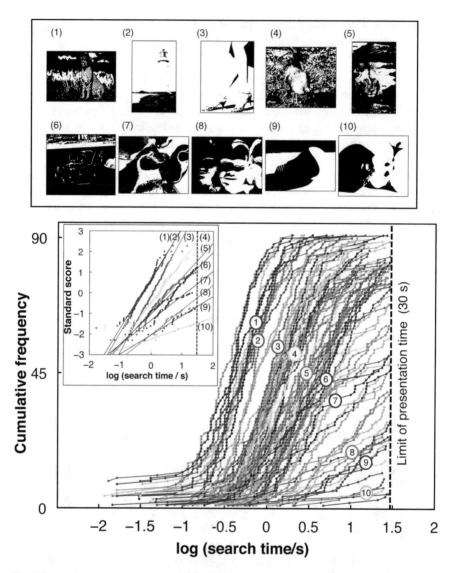

Fig. 2.7 Psychophysical experiment of hidden-figure recognition. Upper panel: Examples of hidden-figure images used in the experiment. Lower panel: Distributions of search times over participants. Each line represents one of the 90 degraded images. The numerical labels from 1 to 10 indicate the example images of the upper panel. Inset: Normal probability plots of logarithmic search times versus standard scores for the example stimuli, each of which is optimally fitted by a line that represents a normal distribution (from [19])

images each of which contained distinct objects that could be verbally described, and the objects were selected from various categories such as animals, transportation, flowers, and buildings (Fig. 2.7: upper panel). We employed 91 participants and measured reaction times of the first correct recognition. The net time needed to recognize the hidden object was obtained by subtracting a reaction time of the undegraded version from one of the degraded versions for each hidden-figure image for each participant. We refer to this difference as a search time. (Reaction times of the undegraded versions were measured after the degraded versions.) Figure 2.7 (bottom panel) shows the cumulative distributions of search times on a logarithmic scale over participants, each line of which represents one of the 90 hidden-figure images. As the examples in the inset show, each distribution well fitted a normal distribution, whose mean m and standard deviation σ were found to have a linear relationship, i.e., $\sigma = Am + B$. For each image, search time of each participant was standardized as z-score, $z = -\frac{\log t - m}{\sigma}$, which indicates relative position of the search time in the distribution. Here the mean m represents the difficulty of the hidden-figure image, and the mean of z values of a participant over images, $\bar{z}$, represents the ability of the participant. Then we have obtained the formulation of search time as follows:

$$\hat{t} = Ce^{\frac{M}{Z}}, \tag{2.1}$$

where $\hat{t}$ is the mean on logarithmic time scale, $\hat{t} = 10^{\overline{\log t}}$, and M and Z, respectively, represent the image difficulty and the participant's ability, being defined by the following transformations:

$$M = \frac{1}{\log e}\left(m + \frac{B}{A}\right), \qquad Z = \frac{1}{1 - A\bar{z}}, \qquad C = 10^{-\frac{B}{A}}.$$

We should note that the formulation (2.1) has the same mathematical structure as the Arrhenius equation, $t = Le^{\frac{E}{kT}}$, which describes time for a chemical reaction to proceed with the parameters of activation energy E, the Boltzmann constant k, temperature T, and a constant L. (The original form of the Arrhenius equation is of the chemical reaction rate, that is, the inverse of the time.) In this analogy, the image difficulty and the participant's ability correspond to the activation energy and the temperature of chemical reaction, respectively. In this sense we may refer to the participant's ability as "cognitive temperature of the brain" of the participant. Since the physical basis of the Arrhenius equation is random collisions of chemical molecules, some stochastic process may underlie the emergent recognition of hidden objects.

We have found another surprising property of image difficulty M, which was periodicity of their distribution of the 90 images as shown in Fig. 2.8. Divided by the period the difficulty could be scaled by a set of natural numbers and their decimal fractions showed a significantly higher concentration around 0 as displayed in the inset, so that the image difficulty can be regarded as a natural number although the observed values included some variation around it. Since the 90 degraded images were made manually using binarization which was set at arbitrary levels for individual images, it is highly unlikely that the degradation itself was done

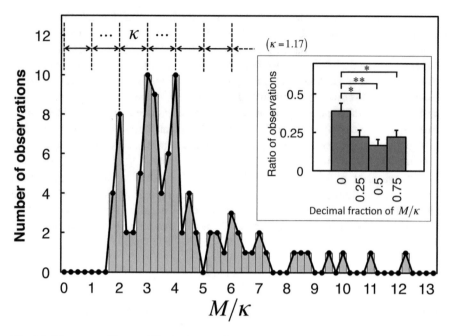

Fig. 2.8 Histogram of image difficulty M scaled by the distribution period κ

selectively in specific discrete levels. This indicates that the search time is given by an exponential function of a natural number. This type of exponential function is given by simultaneous occurrence of multiple discrete stochastic events. For example, the mean number of trials until you hit all the same assigned numbers with a gambling machine having n reels of 10 numbers is 10^n. Since hidden-figure recognition is related to stochastic process as shown above, it seemed hopeful to construct a theoretical model based on Yuragi (stochastic) activity to explain the observed properties of the phenomenon.

2.3.3 Yuragi Model of Hidden-Figure Recognition

To construct a theoretical model, we referred to neurophysiological studies regarding neural representations of visual objects in the brain. Many studies have shown that high-level visual areas (of human and monkey) such as the inferior temporal cortex contain neurons which selectively respond to local features of objects (e.g., a rectangle or a star shape). The studies support the "combination coding" theory in which an object is represented by a combination of simultaneously activated neurons, each of which represents a visual feature of the object [23, 25]. According to this theory, sensory information of a hidden-figure image should fail to activate the object representation because the image degradation eliminated some of local features necessary to achieve the combination of activity for the object representation.

We have constructed the Yuragi model based on the combination coding theory, assuming that neurons in the high-level visual areas exhibit Yuragi activity even in the absence of the selective sensory information [19]. There are many reports of neurophysiology that cortical neurons generate spontaneous firings even in the absence of selective sensory signals and that these firings can be approximated by a Poisson process (i.e., a fundamental stochastic process of independent discrete events) [15, 16]. In our model Yuragi activity occurs stochastically in a Poisson process and coincident activation (i.e., simultaneous activation by chance) of all the neurons that represent missing features of the hidden object completes the combination representing the object to be recognized. The scheme of our Yuragi model is illustrated in Fig. 2.9. The search time of a hidden figure was defined as a time until the coincidence of multiple activation. Let v denote the number of missing features of the object, and let τ denote the time interval within which all the v neurons representing the missing features must become active for their coincident activation to be effective. Let p denote the probability of a Poisson process of Yuragi activation during the interval τ, and the rate of the coincident activation is given by $\frac{p^v}{\tau}$. Thus we have obtained the model value of the search time by

$$\hat{t} = 10^{\overline{\log t}} = \tau e^{-\gamma} e^{-v \ln p}.$$

Comparing this theoretical relation with the experimental formulation (2.1), we have obtained the relations between macroscopic (psychophysical) values and microscopic (of neural model) values as follows:

$$M = \kappa v, \qquad Z = -\frac{\kappa}{\ln p}, \qquad C = \tau e^{-\gamma}, \qquad (2.2)$$

where γ is the Euler's constant ($=0.577$) and κ is a constant that is equivalent to the distribution period of M ($=1.17$). In the third relation, the interval for effective coincident activation is obtained as $\tau = Ce^{\gamma} = 0.082$s. In the second relation, for the participant whose z-score is the group mean ($z = 0$, equivalently $Z = 1$), $p = e^{-\kappa} = 0.310$, so that the mean rate of activity was estimated to be $\frac{p}{\tau} = 3.8$ Hz. It should be noted that Yuragi activity estimated using the experimental data was at such a low rate and seemed to be energy saving.

Using values of v and p determined by the relation (2.2) with the experimental results, we have obtained the consistent results between the model computer simulations of coincident activation (Fig. 2.10) and the experimental results (Fig. 2.7 and Fig. 2.11). The consistent results between the experiment and the Yuragi model strongly support the idea that hidden-figure recognition is based on coincident activation which is characteristically stochastic. In this chapter we showed that Yuragi provides an effective concept to understand flexible energy-saving operations of the brain as well as cellular and molecular-level machinery.

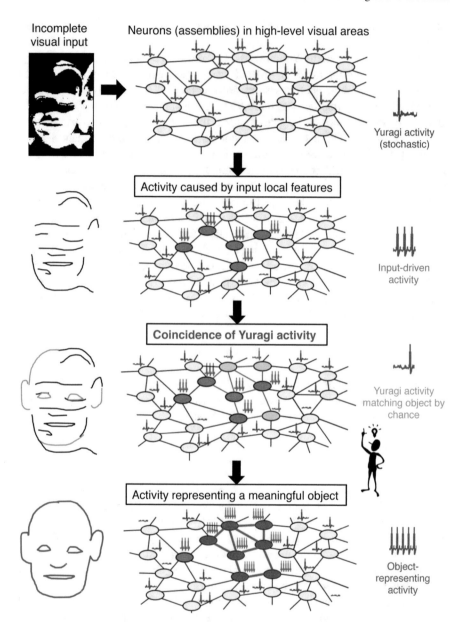

Fig. 2.9 Schematic drawing of our Yuragi model of hidden-figure recognition. First level: Neurons of high-level visual areas generate spontaneous firings (Yuragi activity) before receiving sensory input. Second: An incomplete input of degraded image causes local feature-selective activity. Third: Coincidence of Yuragi activity with the input-driven activity forms the combination coding of the hidden object. Fourth: The hidden figure is recognized in an emergent manner, being separated from the background

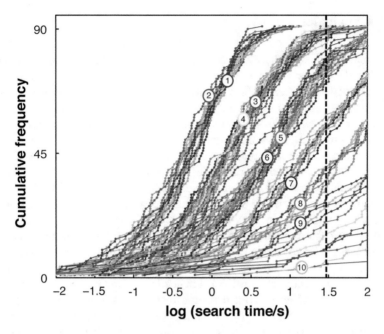

Fig. 2.10 Distributions of logarithmic search times which were obtained by the Monte Carlo simulations of the theoretical model. The numerical labels correspond to Fig. 2.7

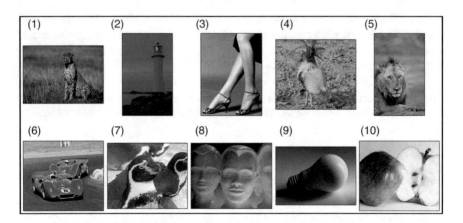

Fig. 2.11 The original images (answers) of Fig. 2.7

2.4 Conclusion

It is commonly thought that Brownian motion and fluctuations are negative factors for artificial machines. They try to block fluctuations and operate at high speed and accuracy controlling their parts precisely. This will work up to some complexity level. However, as the number of parts increases and the complexity increases, the number of combinations of parts to be controlled increases exponentially. As a result, enormous calculation and energy cost for control come to be needed. On the other hand, the muscle, a typical example of biological machines, is packed with hundreds of billions and even trillions of molecular motors. If the parts of muscle are molecular motors and individual molecular motors are precisely controlled like the parts of artificial machine, the muscle would consume a huge amount of energy. As explained in this chapter, however, the simulation using the Yuragi model showed that motions of individual molecular motors look ambiguous but when they assemble to form muscle, this ambiguity enhances their cooperativity in the muscle as a whole, and thus achieves the autonomous response of the muscle to changes in the external world. We obtained the same aspects about the brain function. Activity of an individual neuron of the brain is much slower and less reliable (i.e., more ambiguous) than a device element of the contemporary artificial computer systems. The brain as a system, however, can solve difficult recognition tasks within a reasonable time consuming little energy, which is sometimes very hard and energy consuming for artificial systems. As we showed in this chapter, it is plausible that the brain utilizes stochastic activity (possibly spontaneous activity at the resting level) in order to solve difficult tasks of recognition. To emphasize its positive function, we refer to the stochastic activity as Yuragi activity. We would argue that the Yuragi not only saves energy but also works effectively for flexible and autonomous operation of complex systems.

Acknowledgments We thank Drs. Ben Seymour (Center for Information and Neural Networks) and Lorenzo Marcucci (University of Padova) for critically reading and editing the manuscript.

References

1. Abbondanzieri, E.A., Greenleaf, W.J., Shaevitz, J.W., Landick, R., Block, S.M.: Direct observation of base-pair stepping by RNA polymerase. Nature **438**(7067), 460–465 (2005)
2. Dunn, A.R., Spudich, J.A.: Dynamics of the unbound head during myosin V processive translocation. Nat. Struct. Mol. Biol. **14**(3), 246–248 (2007)
3. Finer, J.T., Simmons, R.M., Spudich, J.A.: Single myosin molecule mechanics: piconewton forces and nanometre steps. Nature **368**(6467), 113–119 (1994)
4. Fujita, K., Ohmachi, M., Ikezaki, K., Yanagida, T., Iwaki, M.: Direct visualization of human myosin ii force generation using DNA origami-based thick filaments. Commun. Biol. **2**(1), 437 (2019)
5. Funatsu, T., Harada, Y., Tokunaga, M., Saito, K., Yanagida, T.: Imaging of single fluorescent molecules and individual ATP turnovers by single myosin molecules in aqueous solution. Nature **374**(6522), 555–559 (1995)

6. Gregorio, G.G., Masureel, M., Hilger, D., Terry, D.S., Juette, M., Zhao, H., Zhou, Z., Perez-Aguilar, J.M., Hauge, M., Mathiasen, S., Javitch, J.A., Weinstein, H., Kobilka, B.K., Blanchard, S.C.: Single-molecule analysis of ligand efficacy in β2AR–G-protein activation. Nature **547**(7661), 68–73 (2017)
7. Hilger, D., Masureel, M., Kobilka, B.K.: Structure and dynamics of GPCR signaling complexes. Nat. Struct. Mol. Biol. **25**(1), 4–12 (2018)
8. Hodges, C., Bintu, L., Lubkowska, L., Kashlev, M., Bustamante, C.: Nucleosomal fluctuations govern the transcription dynamics of RNA polymerase II. Science **325**(5940), 626–628 (2009)
9. Huxley, A.F.: Muscular contraction. J. Physiol. **243**(1), 1–43 (1974)
10. Ishijima, A., Doi, T., Sakurada, K., Yanagida, T.: Sub-piconewton force fluctuations of actomyosin in vitro. Nature **352**(6333), 301–306 (1991)
11. Ishijima, A., Kojima, H., Funatsu, T., Tokunaga, M., Higuchi, H., Tanaka, H., Yanagida, T.: Simultaneous observation of individual ATPase and mechanical events by a single myosin molecule during interaction with actin. Cell **92**(2), 161–171 (1998)
12. Iwaki, M., Iwane, A.H., Shimokawa, T., Cooke, R., Yanagida, T.: Brownian search-and-catch mechanism for myosin-VI steps. Nat. Chem. Biol. **5**(6), 403–405 (2009)
13. Karagiannis, P., Ishii, Y., Yanagida, T.: Molecular machines like myosin use randomness to behave predictably. Chem. Rev. **114**(6), 3318–3334 (2014)
14. Kitamura, K., Tokunaga, M., Iwane, A.H., Yanagida, T.: A single myosin head moves along an actin filament with regular steps of 5.3 nanometres. Nature **397**(6715), 129–134 (1999)
15. Koch, C.: Stochastic models of single cells. In: Biophysics of Computation: Information Processing in Single Neurons Oxford University Press, New York (1999)
16. Lansky, P., Sanda, P., He, J.: The parameters of the stochastic leaky integrate-and-fire neuronal model. J. Comput. Neurosci. **21**(2), 211–223 (2006)
17. Marcucci, L., Yanagida, T.: From single molecule fluctuations to muscle contraction: a Brownian model of A.F. Huxley's hypotheses. PLoS ONE **7**(7), 1–8 (2012)
18. Marcucci, L., Washio, T., Yanagida, T.: Including thermal fluctuations in actomyosin stable states increases the predicted force per motor and macroscopic efficiency in muscle modelling. PLoS Comput. Biol. **12**(9), e1005083–e1005083 (2016)
19. Murata, T., Hamada, T., Shimokawa, T., Tanifuji, M., Yanagida, T.: Stochastic process underlying emergent recognition of visual objects hidden in degraded images. PLoS ONE **9**(12), e115658 (2014)
20. Okada, Y., Hirokawa, N.: A processive single-headed motor: kinesin superfamily protein KIF1A. Science **283**(5405), 1152–1157 (1999)
21. Okada, Y., Higuchi, H., Hirokawa, N.: Processivity of the single-headed kinesin KIF1A through biased binding to tubulin. Nature **424**(6948), 574–577 (2003)
22. Shiroguchi, K., Kinosita, K.: Myosin V walks by lever action and Brownian motion. Science **316**(5828), 1208–1212 (2007)
23. Tanaka, K., Saito, H., Fukada, Y., Moriya, M.: Coding visual images of objects in the inferotemporal cortex of the macaque monkey. J. Neurophysiol. **66**(1), 170–189 (1991)
24. Toyoshima, C., Mizutani, T.: Crystal structure of the calcium pump with a bound ATP analogue. Nature **430**(6999), 529–535 (2004)
25. Tsunoda, K., Yamane, Y., Nishizaki, M., Tanifuji, M.: Complex objects are represented in macaque inferotemporal cortex by the combination of feature columns. Nat. Neurosci. **4**(8), 832–838 (2001)
26. Yanagida, T., Ishii, Y.: Single molecule detection, thermal fluctuation and life. Proc. Jpn. Acad. Ser. B **93**(2), 51–63 (2017). https://doi.org/10.2183/pjab.93.004

Chapter 3
Next-Generation Bio- and Brain-Inspired Networking

Naoki Wakamiya

Abstract The Yuragi model can be regarded as a meta-heuristic algorithm for an optimization problem whose conditions change over time. The activity in the Yuragi equation dynamically strikes a balance between reinforcement learning, which contributes to stability, and random search, which aims to find a solution appropriate for a new condition. Network control methods must provide stable network service by continuously adapting to changing conditions. No information networks are stable; instead, they are exposed to internal and external perturbations, thereby leading the perceived quality and performance to always fluctuate. Using the Yuragi model, a system can adaptively select a state or control option that is appropriate for the current environmental condition, stably remain there as long as the condition does not significantly change, and switch to a new appropriate state once the condition changes. In this chapter, we introduce examples of applications of the Yuragi model to network control.

3.1 Yuragi-Based Routing and Other Network Control Methods

In applying the Yuragi model $\frac{d\mathbf{x}}{dt} = f(\mathbf{x})\alpha + \boldsymbol{\eta}$ to network control, the following steps must be taken.

1. *Associate the attractors (i.e., equilibrium points of vector $\mathbf{x}$) with network control options*
   In the case of adaptive nutrient synthesis of *E. coli* cells, attractors correspond to states with which a cell synthesizes either of two types of nutrients. In the case of network control, vector $\mathbf{x}$ represents the system state. Therefore, we call vector $\mathbf{x}$ the state vector here. Attractors correspond to control options, such as the path used for packet transmission in routing, the sleep/wake state of duty cycling in

N. Wakamiya (✉)
Graduate School of Information Science and Technology, Osaka University, Suita, Osaka, Japan
e-mail: wakamiya@ist.osaka-u.ac.jp

© Springer Nature Singapore Pte Ltd. 2021
M. Murata, K. Leibnitz (eds.), *Fluctuation-Induced Network Control and Learning*,
https://doi.org/10.1007/978-981-33-4976-6_3

a wireless sensor network, and the power mode of a transceiver in a wireless network.

2. *Define activity α as an objective function to be maximized*
   In the case of *E. coli* cells, the activity corresponds to the growth rate. In network control, the activity corresponds to the performance of the system, as it is the optimization target. The one-hop/end-to-end/round-trip delay, throughput, packet loss probability, bit/frame error rate, link utilization, and even user-level metrics, such as Quality of Experience (QoE), can all be used to define the activity depending on the requirements of the system, application, or user. The activity is a monotonically increasing function of the performance measurement and ranges from 0 to 1.

3. *Design function $f(\mathbf{x})$ to have the desired behavior*
   In the case of *E. coli* cells, function $f(\mathbf{x})$ represents the mutually inhibitory relationship between two metabolic pathways synthesizing different nutrients. In applying the Yuragi model to network control, function $f(\mathbf{x})$ must be carefully designed for the system to converge to and remain at the desired attractor when activity α is high and take a random walk when the activity is low. Noise is important for exploring attractors by a random walk and escaping from local optima. However, excessively strong noise threatens the control stability. Furthermore, the desired structure of the state space depends on network control. If distinct and discrete control is required for different environments, such as path selection, the basin of the attractor must be deep to avoid occasionally escaping from a desirable attractor. However, in the case of load distribution, state vector $\mathbf{x}$ can be interpreted as the ratio of tasks to be assigned to servers. Because the ratio is continuous, the state space should be somewhat smooth to allow the system to have intermediate states.

A detailed mechanism is presented below that consists of the following steps.

(a) *Perform network control represented by state vector $\mathbf{x}$*
   In Step 1 above, state vector $\mathbf{x}$ is associated with the system state. It is possible to map each attractor to a control option, where, for example, the system performs control A for an attractor $\mathbf{x}_1$ and B for $\mathbf{x}_2$. However, to take advantage of the Yuragi model, which is in the form of a temporal differential equation and enables continuous adaptation, it is more effective to directly use state vector $\mathbf{x}$. Consider that state vector $\mathbf{x}$ has M elements, where M is the number of control options. Here, we call the elements of a state vector the *state value*, which represents the quality of the corresponding control option. Then, there are two simple alternatives for mapping from the state vector to the control options. The first option is to take the maximum, that is, to adopt the control option with the maximum state value in the state vector. This option is effective in deterministically selecting one option from the alternatives. The second option is to regard the normalized state values as the probability of taking corresponding control options. In the case of load distribution, the probability is regarded as the ratio of tasks to be assigned to servers. It is also possible to select one alternative among several based on the derived probability (i.e.,

stochastic control). The control option with the largest state value is likely to be selected; however, there remains the possibility of selecting another option, which contributes to exploration for an improved solution.

(b) *Evaluate the quality of the current control and derive activity* α

The activity represents the quality of the current control and plays the role of feedback in Yuragi-based adaptive control. A mechanism to obtain measurements is necessary to derive the activity defined in Step 2 above. Depending on the performance measures, either active or passive measurement techniques, or both, are available. In some cases, an additional mechanism is not required to obtain measurement. For example, Transmission Control Protocol (TCP) allows a sender to obtain the round-trip time of a session.

(c) *Update state vector* $\mathbf{x}$ *by the Yuragi model and return to Step a*

State vector $\mathbf{x}$ is updated by substituting the derived activity into the Yuragi equation to reflect the quality of the current control and adapt to the current environmental conditions. A change in state vector $\mathbf{x}$ would cause changes in the network control, such as changing the primary routing path.

The frequency of performing Steps b and c must be determined by taking into account the trade-off between the control adaptability and stability. An appropriate control interval for performing Steps b and c is related to the time scale of changes in the environmental conditions to which the system must adapt. However, if the interval is too short, the system will respond to a small change with high sensitivity. In addition, because it takes time for the system to converge to an attractor, excessively frequent control behaves like random control. Furthermore, especially when an active measurement technique is used to derive the activity, the measurement cost becomes non-negligible. However, excessively slow control delays adaptation, and the control becomes inappropriate for a dynamic environment.

In Sect. 3.2, we extend the Yuragi model to have multiple attractors to apply it to network control. Then, in Sect. 3.3, we provide examples of the application of the Yuragi model to a network control method. In Sect. 3.4, we consider the combination of multiple Yuragi-based control methods, and finally, in Sect. 3.5, we discuss the active exploration of attractors that improves the adaptability of Yuragi-based control.

3.2 Multi-Dimensional Yuragi Model

To apply the Yuragi model to the optimization of network control in a dynamic environment, we must design a function $f(\mathbf{x})$ that defines attractors corresponding to control options, and activity α, which represents the quality of control x in the given environment. The original Yuragi model has only two attractors; that is, $M = 2$, where M is the number of attractors. In general, network control adopts the most appropriate control option among $M \geq 2$ alternatives. For example, each node selects one of its neighbor nodes to forward packets to a specified destination in a multihop network.

We first define a multi-dimensional state vector $\mathbf{x}$ as

$$\mathbf{x} = (m_i \mid i = 1, \ldots, M),$$

where m_i is called the state value, which represents the relative quality of option i among M options. Then, the temporal dynamics of m_i ($1 \leq i \leq M$) is determined by the following stochastic differential equation [6]:

$$\frac{dm_i}{dt} = \frac{s(\alpha)}{1 + \max\limits_{1 \leq j \leq M} m_j^2 - m_i^2} - d(\alpha)m_i + \eta_i. \tag{3.1}$$

Here, α is the *activity*, and $0 \leq \alpha \leq 1$. η is a noise term, and Gaussian white noise with mean 0 and variance 1 is usually adopted.

Equation (3.1) has equilibrium points of the condition

$$\frac{dm_i}{dt} = 0, \ \forall i = 1, \ldots, M.$$

Assuming that m_i is maximal for index $i = k$ without loss of generality, we obtain the following M vectors at the equilibrium points:

$$\mathbf{x}^{(k)} = \left(x_i^{(k)}, \ldots, x_M^{(k)} \right), \ k = 1, \ldots, M$$

with components

$$x_i^{(k)} = \begin{cases} \phi(\alpha), & i = k \\ \frac{1}{2}[\sqrt{4 + \phi(\alpha)^2} - \phi(\alpha)], & i \neq k, \end{cases}$$

where

$$\phi(\alpha) = \frac{s(\alpha)}{d(\alpha)}.$$

Finally, we have M attractors in the form

$$\mathbf{x}^{(1)} = (H, L, \ldots, L)$$
$$\mathbf{x}^{(2)} = (L, H, L \ldots, L)$$
$$\vdots$$
$$\mathbf{x}^{(k)} = (L, L \ldots, H, \ldots, L)$$

$$\vdots$$

$$\mathbf{x}^{(M-1)} = (L, \ldots, H, L)$$
$$\mathbf{x}^{(M)} = (L, \ldots, L, H).$$

In an attractor $\mathbf{x}^k$ ($1 \leq k \leq M$), only the kth entry has a high value, while the other entries have low values. The fact that Eq. (3.1) reinforces state value $m_i = \max_{1 \leq j \leq M} m_j$ by a large activity indicates that an attractor with a larger state value is superior to others. Therefore, the system can accomplish effective control by selecting option k with a high value H ($x_k = H$) at the converged attractor. Furthermore, while the activity is high, the system can stably remain at the effective attractor.

It should be noted that at

$$\phi^* = \frac{1}{\sqrt{2}},$$

there is a special point at which the attractors $\mathbf{x}^{(k)}$ are only defined when $\phi(\alpha) \geq \phi^*$. For $\phi(\alpha) = \phi^*$, we have a single attractor

$$\mathbf{x} = (x_1, \ldots, x_M) \text{ with } x_i = \phi(\alpha), \ \forall i = 1, \ldots, M.$$

In the following applications, we mainly use $s(\alpha)$ and $d(\alpha)$ defined as

$$s(\alpha) = \alpha(\beta \alpha^\gamma + \phi^*)$$

and

$$d(\alpha) = \alpha.$$

Here, parameter $\beta > 0$ is related to the stability of an attractor. With a large β, the difference between the high value $H = \beta + \phi^*$ and the low value $L \approx 0$ increases, and the system state is rarely affected by the noise term η_i when the activity α is high. In addition, parameter $\gamma > 0$ is related to the degree of entrainment of an attractor. With a large γ, the activity α must be sufficiently large to pull the system state to a desirable attractor. Finally, we have

$$\frac{dm_i}{dt} = \frac{\alpha(\beta \alpha^\gamma + \phi^*)}{1 + \max_{1 \leq j \leq M} m_j^2 - m_i^2} - \alpha m_i + \eta_i. \qquad (3.2)$$

An example of the temporal change of state values is presented in Fig. 3.1, where $M = 10$, $\beta = 10$, and $\gamma = 3$. There are 10 fluctuating lines of state values m_i ($1 \leq i \leq 10$) and one thick red line representing the activity α. Initially, the state values are set at random between 0 and 1, and the initial activity α is 0. Therefore,

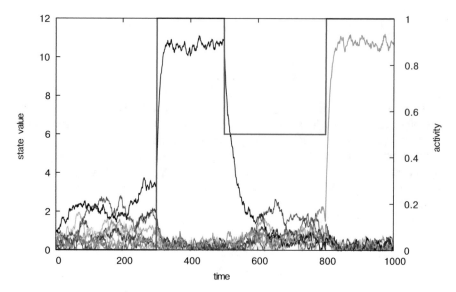

Fig. 3.1 Example of state value dynamics. 10 fluctuating lines represent state values m_i $(1 \leq i \leq 10)$ and one thick red line represents the activity α

the dynamics of the state values is fully governed by the noise term η_i, and the state values change randomly. This corresponds to the random search phase of optimization, where the system selects control options at random.

At time 300, the activity α is manually set to 1.0, signifying that the option with the largest state value (indicated by a black line) is appropriate for the current condition and leads to reasonably high performance. The high activity reinforces the current control by raising the state value m_k, where $k = \text{argmax}_j m_j$, toward the high value H while reducing the other state values to the low value L. By this differentiation, the system stably maintains the current control without being affected by the noise term insofar as the activity remains sufficiently high. Then, at time 500, the activity α is set to 0.5 assuming that there is an unexpected change in the operating condition, which leads the current control to no longer be ideal and decreases the performance. Although the former option k is favored for some time due to possessing a larger state value, it eventually competes with other options. Once the state value m_k begins to be occasionally overtaken by a state value of another option due to the noise term, it begins to decrease by the stochastic differential equation of Eq. (3.2). At time 800, when the state value of the green-colored option is the largest, the activity α is manually set to 1.0. Then, the system converges to a state in which it stably adopts the green-colored option, which is the most appropriate for the new condition.

Next, Fig. 3.2 illustrates how noise influences state values with different β and γ values. The activity α decreases from 1.0 to 0.0 at intervals of 300 units of time. In Fig. 3.2, noise becomes capable of causing a rank change among state values when

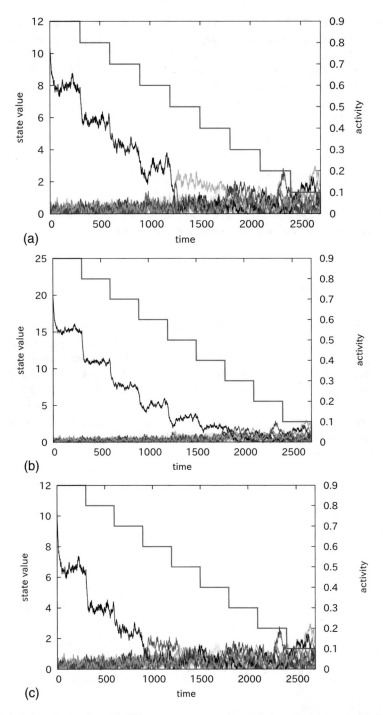

Fig. 3.2 Influence of noise with different parameter settings: (**a**) $\beta = 10$ and $\gamma = 3$, (**b**) $\beta = 20$ and $\gamma = 3$, and (**c**) $\beta = 10$ and $\gamma = 5$

activity $\alpha = 0.5$. At this point, activity α is sufficiently small to allow other state values to exceed the former maximum state value. With a larger β in Fig. 3.2b, the point shifts to approximately $\alpha = 0.3$ because the difference between the high value H and the low value L increases. This indicates that the system does not change its control unless the performance considerably decreases. With a larger γ, the strength of entrainment decreases, and as a result, the point shifts to the left at $\alpha = 0.6$, as illustrated in Fig. 3.2c.

3.3 Application to Single Network Control

In this section, we present examples of applications of the multi-dimensional Yuragi model to a single control method for adaptive, robust, and error-tolerant networking.

3.3.1 Multipath Routing

Routing is the task of finding and maintaining a path to achieve message delivery with high performance. Here, we focus on multipath routing, which is the selection of one path among several candidate paths pre-established between a pair of source and destination nodes [5, 6, 9]. In this case, path selection is performed on a per-session basis at the source node. Path selection must take into account the current environmental conditions, such as the link congestion levels in a wired or overlay network, to achieve the best performance (see Fig. 3.3). A multipath routing mechanism consists of path establishment and path maintenance, which includes path selection [13].

In the path establishment phase, M paths are established between the source node and destination node when the source node has data to send to the destination node and a path has not yet been established between the two nodes. This can be accomplished by a decentralized method similar to the ad hoc on-demand distance vector (AODV) protocol [14], for example. The source node broadcasts route

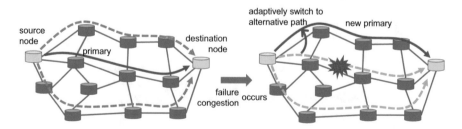

Fig. 3.3 Behavior of multipath routing

request (RREQ) messages with an identical sequence number to that of each of its neighbor nodes. When a neighbor node receives the RREQ message for the first time, it broadcasts the message to all other neighbors. Such rebroadcasting-based message dissemination is called a flooding scheme. If a neighbor node has already received a RREQ message with the same sequence number, it silently discards it to avoid wasting bandwidth by duplicated transmissions. Furthermore, a RREQ message is given a time-to-live (TTL), which decreases by 1 for each broadcasting. When a neighbor node receives a RREQ message with TTL $= 0$, it also discards it to avoid infinite propagation of the message.

When a RREQ message arrives at the destination node or a node that already has a path to the destination, the node first discards the RREQ message and sends a route reply (RREP) message to the source node. The RREP message returns to the source node by taking the reverse path traversed by the corresponding RREQ message. On the way back to the source node, the RREP message sets up the routing information at each intermediate node, as the next-hop node toward the destination is the neighbor node from which the RREQ message arrives at the intermediate node. When the source node receives the first RREP message, it starts using the path in its transmission. The fact that a RREP message arrives first at the destination indicates the quality of the path; therefore, the corresponding path is considered primary. Due to flooding, the source node receives multiple RREP messages taking different paths and takes the first M paths as candidates for multipath routing. From the viewpoint of robustness to node and link failures, it is preferable to have node- or link-disjoint paths as M alternatives. Initially, the primary path is given a high state value, while the other paths are given a low state value. The initial activity α is set to 1.0, signifying that the primary path is the best at the beginning.

The source node performs the above process at the beginning of message transmission to a certain destination node. However, when the number of available paths decreases to M_{min} due to node or link failures, the source node reinitiates the procedure to obtain new M alternative paths.

In the path maintenance phase, the source node performs Yuragi-model-based path selection. At regular intervals, the source node transmits a control message to the destination node to evaluate the quality of the current path. Here, we consider using the one-way delay from the source node to the destination node as the performance measure; however, other metrics, such as throughput and delay jitter, can also be used. By using one-way delay $d(h)$ obtained by the hth control message, the source node first derives the instant activity $\alpha'(h)$ as

$$a'(h) = \frac{\min_{0 \leq k \leq W-1} d(h-k)}{d(h)},$$

where W is the measurement window ($W \geq 2$). If the latest delay $d(h)$ is the minimum of the previous W measurements, the instant activity $\alpha'(h)$ is 1.0, signifying that the current path is optimal for the current environmental conditions.

Then, activity $\alpha(h)$ is updated as

$$\alpha(t) = \begin{cases} \alpha'(h), & \text{if } \alpha(h-1) \le \alpha' \\ \rho\alpha'(h) + (1-\rho)\alpha(h-1), & \text{otherwise} \end{cases}. \tag{3.3}$$

The reason for not directly using $\alpha'(h)$ as α in updating the state values is to avoid *flapping*, where the primary path frequently changes in response to sudden small changes in $\alpha'(h)$. In general, because network bandwidth is shared among multiple source–destination pairs, one-way delay is not constant and continuously fluctuates. If we substitute the instant activity α' into Eq. (3.2), the state values m_i are easily affected by sudden changes in the one-way delay. Consequently, the source node frequently changes its primary path, which leads to unstable communication performance. Furthermore, because changing paths in a single session directly and indirectly affects other sessions, cascading reactions are triggered, and the entire network becomes unstable.

After updating state values m_i, the source node selects the path with the maximum state value as the primary path and uses it in message transmission. However, such a strategy sometimes degrades the adaptive behavior of the system. The system stably remains at an attractor (i.e., path) that was considered optimal in the past unless the activity sufficiently decreases for the noise term to be significant in the Yuragi model. Therefore, even if one of the alternatives can provide better performance, the system is not aware of this and cannot move to the corresponding attractor. One solution is to stochastically select an attractor using

$$p_i = \frac{m_i}{\displaystyle\sum_{1 \le j \le M} m_j}, \tag{3.4}$$

where p_i is the probability of selecting attractor i. However, to accomplish effective control by stochastic attractor selection, the high value H and the low value L should be close to each other. Furthermore, the secondary path is not necessarily better than the primary path. The performance thus deteriorates and becomes unstable. Therefore, there is a trade-off between adaptability and stability. In Sect. 3.5, we discuss the exploration of attractors.

Figure 3.4 presents the results of a simulation. A network of 100 nodes and 219 links was generated using the Waxman model [17], and the capacity and propagation delay of each link were 50 Mbps and 100 ms, respectively. There were a number of continuous-bit-rate sessions that continuously transmitted messages at 8 Mbps. Node pairs were selected such that the total load on the network amounted to approximately 0.5. At time 600 s, a randomly selected 5% of links were simultaneously removed; however, no sessions were completely disconnected. For comparison purposes, we used a simple multipath routing scheme called *Greedy* that had the same set of candidate paths as the Yuragi-based method. This scheme periodically measured the delays of all candidate paths at the same regular intervals as the Yuragi-based method. Then, it selected the path with the minimum delay. For

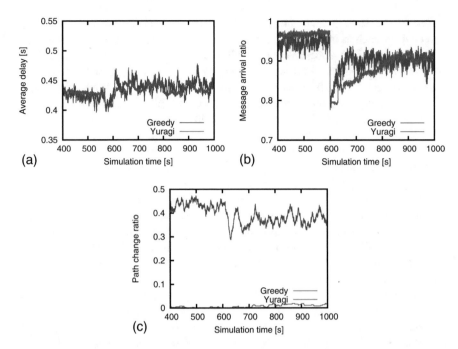

Fig. 3.4 Performance of Yuragi-based multipath routing: (**a**) average delay, (**b**) message delivery ratio, and (**c**) path change ratio

the Yuragi-based method, the parameters were set to $\beta = 10$, $\gamma = 3$, and $\rho = 0.1$, and the number of candidate paths (i.e., M) was 3.

As illustrated in Fig. 3.4a, the average delay was similar for the Yuragi and Greedy methods. However, the Yuragi method achieved more stable communication with less fluctuating delays than the Greedy method. In Fig. 3.4b, the Yuragi method initially achieved a higher message delivery ratio than the Greedy method. When failures occurred, the message delivery ratio decreased for both the Greedy and Yuragi methods. Because the Greedy method examined all candidate paths, it was able to recover from failures faster than the Yuragi method. However, the message delivery ratio after recovery was higher in the Yuragi method. The fast recovery of the Greedy method was achieved at the cost of stability. Figure 3.4c indicates that the path change ratio, which is the ratio of sessions changing the primary path, was much higher in the Greedy method than in the Yuragi method.

3.3.2   Routing in Mobile Ad Hoc Networks

Here, we consider routing in mobile ad hoc networks (MANETs) as another application of the multi-dimensional Yuragi model. In a MANET, there is no fixed infrastructure, such as routers in a wired network. Instead, the network consists of end devices, such as laptops and mobile terminals. There are also no physical links, such as optical fibers, and a link in a MANET is virtual. A pair of nodes is considered to have a link when the nodes are within the wireless signal range and can send and receive packets to and from each other. As a result, routing in MANETs is not a trivial task. Nodes can move, and wireless communication is unstable; thus, a link is temporary and unreliable. Furthermore, unlike a wired link, a wireless link is not necessarily bidirectional because it is not guaranteed that a pair of nodes can transmit and receive packets to and from each other. The nodes themselves are not reliable because they are the equipment of users and are thus more fragile than routers. For example, they may be suddenly turned off by users.

Therefore, routing in MANETs does not involve the selection of one path by relying on the sustained existence of paths for a certain duration of time. Instead, it involves the selection of a neighbor node as a next-hop node expecting that a packet reaches the destination by being relayed by other nodes. Considering the aforementioned challenges, Yuragi-based control is promising and feasible due to its adaptability and robustness [1, 7, 8].

Similar to Yuragi-based multipath routing in wired networks, an initial path is established by flooding RREQ packets [1]. However, unlike multipath routing in which routing information is managed on a per-session basis, all intermediate nodes maintain routing information on a per-destination basis. Specifically, route information for a certain destination consists of a state vector $\mathbf{m} = (m_1, \ldots, m_M)$, where M is the number of neighbors, and activity α. A node maintains a list of neighbor nodes by exchanging HELLO packets, as in AODV. Here, neighbor nodes are nodes that are within the wireless communication range and can exchange packets with each other (i.e., bidirectional links). When a node receives a RREP packet for the first time for a certain destination, it sets the corresponding activity to 1, indicating the availability of a route to the destination.

The source node transmits a packet to a neighbor node with the maximum state value. Each intermediate node forwards the received packet to the designated destination node by selecting a next-hop node in the same manner (see Fig. 3.5a). During route establishment or exploration, state values can be close to each other. In this case, due to random next-hop selection, a node occasionally receives a packet to a destination for which it does not have any routing information. It then sets up new route information with initial random state values. Consequently, a packet is forward to a randomly selected neighbor node due to the effect of noise.

When the destination node receives a packet, it generates a feedback packet and sends it back to the source node. The feedback packet returns to the source node by being relayed by the same intermediate nodes of the received packet. At the same time, it calculates the number of hops from the destination node to each intermediate node, which is used by intermediate nodes and the source node to update the activity for the destination (see Fig. 3.5b). In forwarding a feedback packet, an intermediate node uses broadcasting, which allows neighbor nodes along a path that do not directly participate in data forwarding to update their routing information for the destination node. In addition, by receiving a feedback packet,

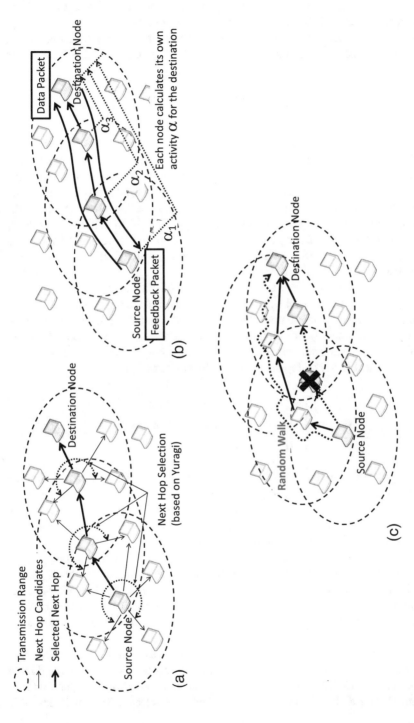

Fig. 3.5 Overview of Yuragi-based mobile ad hoc network (MANET) routing: (**a**) next-hop selection, (**b**) activity calculation, and (**c**) route recovery

neighbor nodes can update the list of neighbor nodes without exchanging HELLO messages. This contributes to a reduction in control overhead.

When a node receives a feedback packet at time t, it evaluates the activity using the number of hops $d(t)$ to the destination node embedded in the feedback packet as

$$\alpha(t) = \frac{\min_{0 \le k \le T} d(t - k)}{d(t)}, \qquad (3.5)$$

where $W > 0$ is the measurement window. If the number of hops increases, this signifies that the current route to the destination node is no longer appropriate and attempts to determine that a superior route should be made. Therefore, the activity decreases by Eq. (3.5), and the noise induces a random walk. The reason for not using the moving average, as in Yuragi-based multipath routing, is that the environmental changes are faster and more significant in MANETs. Smoothing, such as in Eq. (3.3), delays reactive control, such as of the movement of nodes and thus reduces the adaptability of Yuragi-based routing, which results in performance deterioration and even route disconnection.

In addition, to handle the instability of environmental conditions, Yuragi-based MANET routing has a mechanism to initialize the information of a route that is no longer available. When a node does not receive any feedback packets from the destination node due to, for example, link disconnection, it loses the opportunity to update the routing information. To allow the system to escape the stalling condition, the activity decays over interval τ by the following equation:

$$\alpha(t) = \begin{cases} \alpha(t_0), & \text{if } t - \tau \le k\tau < t_0 < t \\ \alpha(t_0) - \delta, & \text{if } t - \tau \le t_0 \le k\tau < t, \\ \alpha(t - \tau) - \delta, & \text{otherwise} \end{cases}$$

where t_0 is the time at which the most recent feedback packet is received, and δ is the decay constant. The activity decay mechanism is performed regardless of feedback packet reception. Therefore, when a feedback packet does not arrive, the activity continuously decreases. Consequently, the state values become random, and next-hop selection returns to the random walk phase.

In MANETs, a route can be easily disconnected for a variety of reasons, such as node movement and wireless interference. Conventional routing protocols require a designated route recovery mechanism; however, Yuragi-based MANET routing is inherently robust to route disconnection. Packets occasionally take a random walk seeking a new route to the destination node due to the noise term and random state values (see Fig. 3.5c).

Figure 3.6 presents the performance of Yuragi-based MANET routing in comparison with AODV. A total of 256 nodes were randomly but evenly distributed in an area of $1500 \times 1500 \, \text{m}^2$, and the average communication range was approximately $510 \, \text{m}$. During a simulation of $1000 \, \text{s}$, a randomly selected 25% of nodes were

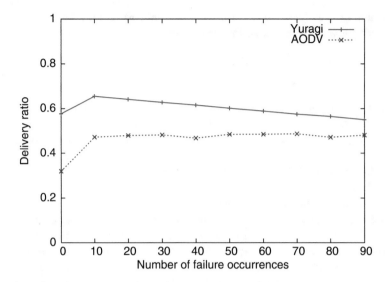

Fig. 3.6 Performance of Yuragi-based mobile ad hoc network (MANET) routing

removed from the network, simulating a sudden switch off. After a certain amount of time, which was determined by dividing the simulation time by the number of failure occurrences on the x-axis, the nodes resumed operation and returned to the network, simulating the switch-on condition, and a newly selected 25% of nodes were removed. Therefore, a larger number of failure occurrences corresponded to the frequent switching on/off or joining/leaving of nodes. When a node was removed, all links from the removed node to neighbor nodes were lost. Thus, routes were very likely to be disconnected. However, when a node rejoined the network, new links became available, indicating that there may appear a better route than the current one. The occasional random walk explained in the previous paragraph thus contributes to finding an improved route. As illustrated in Fig. 2.6, independent of the number of failure occurrences, Yuragi-based MANET routing achieved a higher delivery ratio than AODV in a dynamic environment.

For other types of applications, the reader is referred to coverage control in a wireless sensor network [3] and mobility management in a mobile core network [18].

3.4 Application to Multi-Network Control

Thus far, we have introduced examples of applications of the Yuragi model to single network control. In this section, we consider scenarios in which multiple Yuragi-based network control methods are incorporated with each other to accomplish a system-wide goal.

Assuming that there are C Yuragi-based network control methods, a naive application is to consider the following equation:

$$\frac{d\mathbf{x}_i}{dt} = f_i(\mathbf{x}_i)\alpha_i + \boldsymbol{\eta}_i. \tag{3.6}$$

Method i attempts to maximize its own activity α_i by adaptively selecting the most appropriate state $\mathbf{x}_i$ through independent noise $\boldsymbol{\eta}_i$. For example, routing in the network layer finds, establishes, and maintains the best path from a source node to a destination node, where the activity is derived from the delay. In addition, flow control in the transport layer effectively uses the available bandwidth of the established path by using the activity defined based on the throughput. Routing and flow control generally behave independently of each other; however, due to the layered network control architecture, they inherently interact with each other. As indicated in Eq. (3.6), control methods attempt to maximize their own activity in an independent and selfish manner; therefore, maximization of global performance is not always guaranteed. It should be noted that the control entities are not necessarily on the same network equipment. In the above case, routing is performed by both end hosts and routers; however, flow control is performed by end hosts alone.

System-level optimization can be achieved by introducing more direct interaction between Yuragi-based control methods. There are various formulations. For example,

$$\frac{d\mathbf{x}_i}{dt} = f_i(\mathbf{x}_i)\alpha + \boldsymbol{\eta}_i \tag{3.7}$$

signifies that all C control methods share and maximize the common global activity α. Although such Yuragi model-based control is useful, careful design is required to avoid harmful interaction, where methods degrade each other's performance similarly to the above naive application. The behavior of a control method to improve global activity sometimes disturbs other control methods. The disturbed methods then apply control that is considered necessary to improve the global activity, which further affects other control methods. Consequently, the system states fluctuate considerably, and globally optimal control cannot be expected. A mechanism is also necessary to share the global activity α among control entities without loss of consistency.

A more general Yuragi model with shared activity is formulated as

$$\frac{d\mathbf{x}_i}{dt} = f_i(\mathbf{x}_i)A_i(\boldsymbol{\alpha}) + \boldsymbol{\eta}_i \ ,$$

where control method i takes into account the quality of other control methods by function $A_i(\boldsymbol{\alpha})$, where $\boldsymbol{\alpha} = (\alpha_1, \ldots, \alpha_C)$. Therefore, this method can be regarded as a heuristic algorithm for a multiobjective optimization problem.

Another direction to achieve explicit interaction among control methods is to use

$$\frac{d\mathbf{x}_i}{dt} = f_i(\mathbf{x})\alpha_i + \eta_i \ ,$$

where $\mathbf{x} = (\mathbf{x}_1, \ldots, \mathbf{x}_C)$, signifying that control method i takes into account the states of other control methods. The control methods are more tightly coupled with each other in this case, and finer cooperative control can be achieved. However, the control methods must exchange and share information about their internal state values. In addition, the dimension of the state space increases considerably. As a result, the adaptive and stable selection of an effective attractor becomes difficult.

3.4.1 Multipath Routing in Layered Networks

The first scenario of multi-network control is multipath routing in layered networks, which is an extension of Yuragi-based multipath routing discussed in Sect. 3.3 [15, 16]. Figure 3.7 illustrates a targeted system, where we assume that network virtualization technology, such as software-defined networking (SDN) or network function virtualization (NFV), is used. Network virtualization enables the construction of virtual networks over physical networks, with which application- or

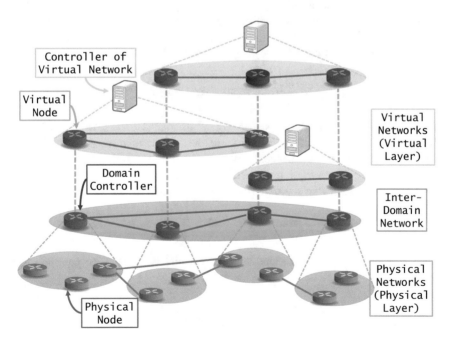

Fig. 3.7 Layered network

service-specific network functions are delivered. Unlike overlay networking, virtual nodes and virtual links appear as physical entities dedicated to a tailored virtual network. Virtual networks can adopt independent network functions even at lower layers of network protocols.

In Fig. 3.7, the global physical network at the bottom consists of physical domain networks composed of physical nodes (e.g., routers or switches) and physical links (e.g., optical fibers). For network virtualization, each physical domain network is managed by a controller. The controller performs resource management to share and assign physical network resources, such as the computational/memory capacity of nodes and bandwidth of links, to virtual networks. It also performs other network control, such as routing. Controllers constitute an inter-domain virtual network in which virtual links correspond to inter-domain physical links between border nodes (i.e., physical nodes at the border of domain networks). Furthermore, by virtualizing domain controllers and inter-domain links, several application- or service-oriented virtual networks are constructed above the inter-domain virtual network. For the purpose of network management, each virtual network has a controller. Due to resource limitations, it is generally impossible to allocate dedicated and independent resources to each virtual network. Instead, for the efficient use of resources, resources are shared among virtual networks. Therefore, when a session in a virtual network changes its path, not only other sessions in the same virtual network, but also sessions in other virtual networks, are affected, and in the worst case, the cascading effect is triggered on a system-wide level.

Each of the virtual nodes s performs multipath routing to each of the destination nodes d in the same virtual network by using state vector $\mathbf{m}_{s,d} = (m_{s,d,1}, \ldots, m_{s,d,M_{s,d}})$, where $M_{s,d}$ is the number of candidate paths from a source node s to a destination node d, activity $\alpha_{s,d}$, and the Yuragi equation

$$\frac{d\mathbf{m}_{s,d}}{dt} = f(\mathbf{m}_{s,d})\alpha_{s,d} + \eta_{s,d}. \tag{3.8}$$

We use Eq. (3.2) to update the state values; however, for the sake of simplicity, we use the general form of the Yuragi model here. At regular intervals of T s, the source node emits a control packet to measure the one-way end-to-end delay to the destination node. Upon receiving a feedback packet from the destination node for the hth control packet, the smoothed delay $\bar{d}_{s,d}(h)$ is derived as

$$\bar{d}_{s,d}(h) = \rho \min_{0 \leq k \leq W} d_{s,d}(h - k) + (1 - \rho)\bar{d}_{s,d}(h - 1),$$

where ρ ($0 < \rho < 1$) is the smoothing factor and $W > 0$ is the measurement window. Then, the activity is updated as

$$\alpha_{s,d} = \frac{\bar{d}_{s,d}(h)}{d_{s,d}(h)}.$$

Similarly, a physical node x performs Yuragi-based multipath routing to a destination node y in the same physical domain network by using the state vector $\mathbf{m}_{x,y} = (m_{x,y,1}, \ldots, m_{x,y,M_{x,y}})$, activity $\alpha_{x,y}$ (which is defined similarly to $\alpha_{s,d}$), and the equation

$$\frac{d\mathbf{m}_{x,y}}{dt} = f(\mathbf{m}_{x,y})\alpha_{x,y} + \eta_{x,y}. \tag{3.9}$$

Because virtual nodes and physical nodes use their own state values and activity, we call this method *Independent*.

For cooperative routing, we consider three other methods: *BottomUp*, *TopDown*, and *Both*. In BottomUp, a virtual node s takes into account the quality of routing in the corresponding physical domain network. In place of the activity $\alpha_{s,d}$ in Eq. (3.8), the following activity $A_{s,d}$ is used:

$$A_{s,d} = \alpha_{s,d} \frac{\sum_{x,y} \alpha_{x,y}}{N_s(N_s - 1)}. \tag{3.10}$$

Here, the denominator $\sum_{x,y} \alpha_{x,y}$ is the sum of the activities of all pairs of physical nodes in a physical domain network s corresponding to a virtual node s. N_s is the number of physical nodes in the domain network s. Therefore, the activity $A_{s,d}$ reflects both the quality of virtual layer routing and the average quality of physical layer routing. The selection of paths at a virtual node corresponds to the selection of border nodes in the corresponding physical domain network. All traffic for a virtual node d (i.e., physical domain network d) generated in the physical network s is destined for the selected border node. Therefore, routing in a virtual layer affects the performance of intra-domain routing. Using Eq. (3.10), we can achieve lower-layer-aware routing, and the performance of physical layer routing is improved. Consequently, the performance of virtual layer routing increases because the end-to-end delay consists of both the intra-domain and inter-domain transmission delays.

In contrast, a physical node uses the following activity $A_{x,y}$ in TopDown:

$$A_{x,y} = \alpha_{x,y} \frac{\sum_v \frac{\sum_d \alpha_{s,d}}{N_v(N_v-1)}}{V_s}. \tag{3.11}$$

Here, s is a virtual node of physical domain network s in which physical node x resides. Denominator $\sum_s \frac{\sum_d \alpha_{s,d}}{N_v-1}$ is the sum of the average activities $\alpha_{s,d}$ in the virtual network v to which virtual node s belongs. N_v is the number of virtual nodes in virtual network v, and V_s is the number of virtual networks to which virtual node s belongs. A virtual node can be a member of multiple virtual networks. For example, the rightmost virtual node in Fig. 3.7 is a member of two virtual networks. As discussed, routing in a physical domain network affects the performance of packet delivery to border nodes. Thus, TopDown favors upper-layer routing.

Finally, Both is a combination of BottomUp and TopDown. A virtual node uses Eq. (3.10), and a physical node uses Eq. (3.11) as the activity in updating the state values. Virtual nodes and physical nodes cooperatively aim at minimizing the one-way end-to-end delay in both the virtual and physical layers simultaneously. Although system-level optimization is expected, there is a risk of routing instability due to the mutual and potentially harmful interaction between layers.

In Fig. 3.8, the results of a simulation are presented. A random physical network consisted of 100 domain networks, and each domain network consisted of 100 physical nodes. These networks were generated using the Waxman model. Three random virtual networks were overlaid on the physical network, each of which consisted of 50 virtual nodes. These virtual networks were subgraphs of a global inter-domain network of 100 nodes (i.e., domains). All pairs of virtual nodes had a traffic demand of 1 unit. Thus, there were $50 \times 49 \times 3 = 7350$ sessions in a virtual layer. Furthermore, inter-domain traffic was distributed among physical nodes with additional traffic, such that the ratio of traffic generated by a single physical node was inter-domain : intra-domain $= 8 : 2$. The capacity was set to 10,000 units for each inter-domain link and randomly selected from 40,000, 20,000, 10,000, and 5000 units for intra-domain links. We called the sessions using bottleneck links that accommodated the largest number of shortest paths between pairs of virtual nodes *bottleneck sessions*. At 100 s, the traffic of the bottleneck sessions was increased to 2 units to cause congestion. At 2000 s, we further caused a second and drastic change

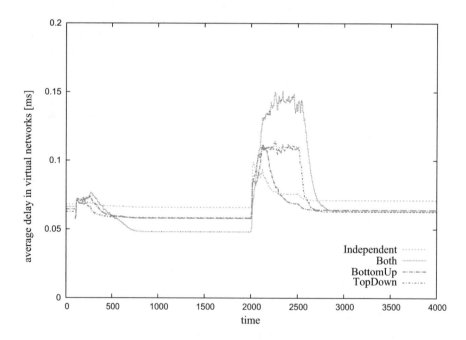

Fig. 3.8 Average delay in virtual networks

by reducing the traffic of the bottleneck sessions to 1 and increasing the traffic of the other sessions to 2. For the Yuragi-model-based routing parameters, we used $\beta = 20$, $\gamma = 5$, $\rho = 0.5$, and $W = 50$. Each of the virtual and physical sessions had at most three candidate paths ($M = 3$). Initially, all sessions used the shortest path among the candidates independently of the methods.

As illustrated in Fig. 3.8, when the traffic demand changed, the average delay instantaneously increased independently of the methods. However, as time passed, the average delay decreased and reached a stable condition (i.e., stable attractor). The Independent method converged the fastest among the four methods. This indicated that independent adaptive control was simpler than the other methods, and competition was successfully mediated by Yuragi-based stochastic control. However, the obtained solution was not optimal, as the average delay was larger than that of the other methods. In contrast, explicit inter-layer interaction led to better performance in Both, BottomUp, and TopDown. In particular, Both achieved the best performance; however, it took the largest amount of time to adapt to the changes.

3.4.2   Cluster-Based Routing in Wireless Sensor Networks

The second scenario in which multiple Yuragi-based network control methods are incorporated with each other is a combination of clustering and routing in wireless sensor networks (WSNs). A WSN consists of small devices equipped with one or more sensors, a processor, a wireless transceiver, and a battery. Each device monitors the surroundings and reports the sensed data to another wireless device called a base station, or sink, using wireless communication. The sink processes the received data and/or sends them to an external server via a wired or wireless network. Owing to the simplicity and ease of deployment of WSNs, a variety of applications have been developed and are available on the market, such as automatic meter reading, factory monitoring, and environmental surveillance. As a result, WSNs have become key technologies in the era of Internet of Things (IoT), in which an enormous number and variety of devices are connected to the Internet, and big data is gathered, analyzed, and utilized to promote safety and comfort in society.

A WSN generally has a tree topology whose root is a sink and whose nodes and leaves correspond to sensor nodes, as illustrated in Fig. 3.9a. The majority of communication in a WSN is data gathering, where sensed data are sent from nodes to a sink. Each sensor node appoints a neighbor node that is closer to the sink as the parent node. The distance from the sink is defined as the number of hops from the sink rather than the Euclidean distance. Data messages are then sent to the parent node, which further forwards the received messages to its parent node. By being relayed by ancestors, messages reach the sink.

For decades, energy savings have been the primary concern in developing and operating WSNs. For example, in a power plant, a WSN is expected to operate without maintenance for at least 10 years. In addition to miniaturization, weight

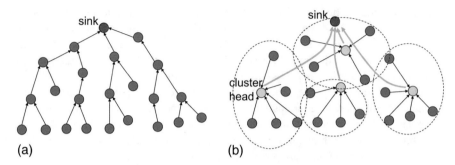

Fig. 3.9 Data gathering in a wireless sensor network: (**a**) tree-based data gathering and (**b**) cluster-based data gathering

savings, and power savings of circuitry and devices, much effort has been devoted to achieve energy-efficient communication. A sensor node consumes energy in transmitting and receiving messages using a wireless transceiver; thus, reducing the number of transmissions and receptions as well as the transmission power is important. Clustering is a technique used for this purpose. In clustering, nodes are first grouped into clusters. One node in each cluster becomes a cluster head (CH) and is responsible for collecting sensor data from the other nodes, called cluster members (CMs), in the same cluster by intra-cluster routing and for transmitting aggregated data to a sink by inter-cluster routing. Sensor nodes in tree-based data gathering consume energy in the reception and transmission of messages; however, in a clustered WSN, the CH performs most of the tasks, and members only send their data to the CH.

Because nodes appointed as the CH have a high burden, a mechanism to distribute or rotate the role of the CH among nodes is required to balance the energy consumption among nodes and prolong the lifetime of the WSN. Selecting a node with the maximum residual energy in the vicinity is simple and somewhat effective. However, such greedy selection often converges to local optima, and global optimization leading to the maximum lifetime of the WSN is not achieved. In addition, the cluster structure continuously changes because the residual energy changes over time. As a result, inter-cluster routing becomes unstable and continuously searches for suitable paths.

If a central controller (e.g., a sink) has complete knowledge of the entire system, an optimal schedule to select nodes to appoint CHs can be derived. However, information such as the amount of residual energy, the neighborhood relationship, and the amount of data to send for optimization is not necessarily available at the sink, as it changes depending on uncontrollable and unexpected conditions. The neighborhood relationship is not stable due to the unstable and unreliable wireless communication, and the amount of data to send is also not constant in some applications. For example, if an irregular value is obtained by a sensor, additional data will be collected by multiple measurements or by additional sensors, and the amount of data to send will increase. As a result, energy consumption will also change, and the amount of residual energy will not decrease linearly.

Furthermore, we must consider a method for collecting aggregated data from CHs, that is, inter-cluster routing, even if we can assume that CMs directly send their data to a CH owing to their proximity. In inter-cluster routing, each CH selects a node to send aggregated data toward a sink among nodes in the wireless communication range. Here, we also require an effective and efficient mechanism of next-hop selection. A simple algorithm involves selecting the node with the minimum number of hops from the sink and with the maximum residual energy in the vicinity. However, there is a possibility that a long, unreliable, and even disconnected path will be constructed, thereby reducing the probability of successful data gathering. Furthermore, such an algorithm can cause significant congestion due to traffic concentration on the shortest and most energy-rich paths, thereby reducing the lifetime of the WSN. Centralized control is thus not feasible for inter-cluster routing. Therefore, an autonomous, distributed, and adaptive mechanism is necessary for both clustering and routing.

Application of the Yuragi model is a feasible solution to the combination of these challenges. A layered model of clustering and routing is considered, as illustrated in Fig. 3.10 [11]. In the clustering layer, nodes aim to balance energy consumption by autonomously appointing a node with large residual energy as the CH in the vicinity. In the routing layer, CHs aim to perform efficient and reliable data gathering by minimizing the data gathering delay. As illustrated in the figure, an aggregated data message of the CH is first sent to a gateway (GW) node, which is a CM residing in an overlapping area with a neighboring cluster. A GW node can reach both CHs.

The Yuragi-based clustering and data gathering method consists of a clustering phase, a data gathering phase, and a routing phase. The clustering phase is repeated at regular intervals of $T_{clustering}$, which is synchronized among all nodes. During the clustering phase, each node determines its state as either CH, GW, or CM. Then, at regular intervals of $T_{data} < T_{clustering}$, sensor data of GWs and CMs are gathered at a designated CH. After $T_{aggregate} < T_{clustering}$, during which the received data are deposited and aggregated by a CH, all CHs and GWs move to the routing phase. Each CH and GW selects a next-hop node from neighboring GWs and CHs, respectively, to send aggregated data to the sink.

At the beginning of the clustering phase, all nodes are in the NULL state. First, the sink broadcasts a cluster head claim (CHC) message to all neighbors and becomes a CH. Hearing the CHC message, nodes in the vicinity become CMs of the sink. Next, those CMs broadcast a cluster formation (CF) message to their neighbors. Each node i receiving a CF message first evaluates the quality of the current cluster structure by computing the clustering activity α_i^c by the following equation:

$$\alpha_i^c(h) = \rho_c \alpha_i^c(h-1) + (1 - \rho_c) \frac{\min_{k \in CH_i} r_k}{\max_{j \in N_i} r_j}.$$

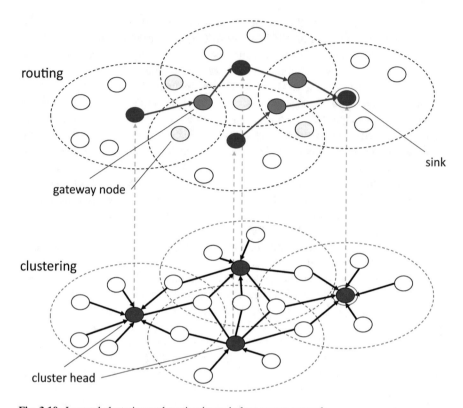

Fig. 3.10 Layered clustering and routing in a wireless sensor network

Here, ρ_c is the smoothing factor, and N_i is the set of neighbor nodes of node i including node i itself. CH_i is a set of CHs from which node i receives a CHC message, including node i itself, and r_k is the amount of residual energy of node k. Therefore, the clustering activity reflects the ratio of the minimum residual energy of nearby CHs to the maximum residual energy of nodes in the vicinity.

Next, node i updates the state vector $\mathbf{m}_i^c = \left(m_{i,1}^c, \ldots, m_{i,|N_i|}^c \right)$. The state value $m_{i,j}^c$ indicates the quality of node $j \in N_i$ as a CH. $m_{i,1}^c$ corresponds to node i itself. We also use Eq. (3.2) to update the state vector $\mathbf{m}_i^c$ by substituting $\alpha_i^c(h)$; however, we replace η_i with $(1 - \alpha_i^c)\,\eta_{i,j}$ to have more stable clustering control. While the activity is relatively high (not necessarily close to 1), the influence of the noise term remains sufficiently small not to initiate a random walk search. However, when the activity becomes very small, indicating that the current cluster structure is no longer appropriate, another cluster structure is examined to adapt to the new conditions.

Then, upon receiving a CF message, node i in the NULL state sets its backoff timer t_i as

$$t_i(h) = T_{backoff} \left(1 - \frac{\min \left(\max_j m_{i,j}(h), m_{i,1}(h) \right)}{\max_j m_{i,j}(h)} \right)^2. \qquad (3.12)$$

Here, $T_{backoff}$ is the maximum waiting time. When the backoff timer expires without receiving any CHC messages from neighbors, node i broadcasts a CHC message, moves to the CH state, and becomes a CH. Equation (3.12) expresses that the more residual energy is in the vicinity, the shorter the backoff timer is. Therefore, a node with more residual energy is more likely to broadcast a CHC message and become a CH. When a node in the NULL state receives a CHC message independently of setting of a backoff timer, it cancels the timer, becomes a CM of the CH from which it first receives the CHC message, and broadcasts a CF message.

When a CM receives multiple CHC messages, it becomes a GW by setting its state to GW. A GW mediates inter-cluster message forwarding and broadcasts a gateway claim (GWC) message to inform nearby CHs of its existence. By receiving GWC messages, a CH obtains a list of nearby GWs. In this way, cluster formation propagates from the sink to the edge of the WSN. Eventually, all nodes determine their state and obtain a list of CHs and GWs.

At the beginning of the hth routing phase, node i, which is either a CH or GW, first evaluates the routing activity α_{ir} by the following equation:

$$\alpha_{ir}(h) = \rho_r \alpha_{ir}(h-1) + (1 - \rho_r) \frac{d_{min}}{d_{latest}}. \qquad (3.13)$$

Here, ρ_r is the smoothing factor, d_{min} is the minimum data gathering delay since the last clustering, and d_{latest} is the latest delay. For a CH or GW to evaluate the routing activity, the sink sends a feedback message by flooding at a regular interval $T_{feedback} < T_{data}$ from the beginning of the data gathering phase. The feedback message contains a list of the reception time of all data messages received in the previous period of $T_{feedback}$. Similar to clustering, using the routing activity α_{ir}, a CH or a GW i updates another state vector for routing, which consists of $\mathbf{m}_i^r = \left(m_{i,2}^r, \ldots, m_{i,|N_i|}^r \right)$. The state value $m_{i,j}^r$ indicates the quality of node $j \in N_i$ as a next-hop node in inter-cluster routing. A CH or GW selects a GW or CH with the maximum state value as the next-hop node and sends the aggregated data.

The two control methods influence each other. Yuragi-based clustering defines the activity as the relative residual energy based on which a backoff timer is set. It aims at balancing energy consumption by rotating the role of CH. In contrast, Yuragi-based routing defines the activity based on the delay in order to achieve fast and reliable data gathering. Yuragi-based clustering affects Yuragi-based routing by changing the cluster structure. At the same time, Yuragi-based routing affects Yuragi-based clustering by consuming energy in data gathering.

We now consider a combination of the two control methods. There are three alternatives: routing-aware clustering, clustering-aware routing, and mutual incorporation. Each node takes into account the routing activity in the derivation of the clustering activity in routing-aware clustering. Therefore, clusters are organized to decrease the delay at the cost of the system lifetime. In contrast, clustering-aware routing takes into account the clustering activity in the derivation of the routing activity. Thus, a next-hop node is selected to balance the energy consumption of nodes; however, such selection worsens the delay and reliability of data gathering.

Here, we present results of routing-aware clustering. Freedom of selection is more limited in inter-cluster routing than clustering. Only CHs and GWs are involved in inter-cluster routing, and selection is only among neighboring GWs and CHs. In the Coupled method, the clustering activity α_{ic} is combined with the routing activity α_{ir} as

$$\alpha_{icr}(h) = \omega\alpha_{ic}(h) + (1 - \omega)\alpha_{ir}(h) ,$$

where ω $(0 < \omega < 1)$ is a weight to balance both activity values.

Figure 3.11 presents the simulation results. We randomly and evenly distributed 81 nodes in an area of $2000 \times 2000\,\text{m}^2$. The node at the center was the sink, and the wireless communication range was set to $250\,\text{m}$. The control parameters were set as follows: $T_{clustering} = 200\,\text{s}$, $T_{data} = 5\,\text{s}$, $T_{aggregate} = 0.4\,\text{s}$, $T_{backoff} = 0.1\,\text{s}$, $T_{feedback} = 2\,\text{s}$, $\rho_c = 0.6$, $\rho_r = 0.8$, and $\omega = 0.5$. The Yuragi parameters were set to $\beta = 20$ and $\gamma = 10$. The number of neighbors of a node was approximately 5 $(M = 5)$ on average. At $3000\,\text{s}$, we generated background traffic to cause environmental changes and evaluate the adaptability. The background traffic generated a message each $0.01\,\text{s}$ from one corner of the area to the other corner. As displayed in Fig. 3.11a, both the Independent and Coupled methods achieved the same data gathering ratio. However, Fig. 3.11b reveals that the Coupled method produced a shorter one-way delay than the Independent method, especially after the introduction of background traffic. This implies that the clusters were organized so that aggregated data could detour around the congested area. To evaluate the balance of energy consumption, we used the fairness index F, defined as

$$F = \frac{\left(\sum_{i=1}^{N} r_i\right)^2}{\left(N \sum_{i=1}^{N} r_i^2\right)} ,$$

where N is the number of nodes. The fairness index was 0.8953 for Independent and 0.8931 for Coupled. Thus, faster inter-cluster routing was accomplished without sacrificing either reliability or lifetime by the Yuragi-based routing-aware clustering method.

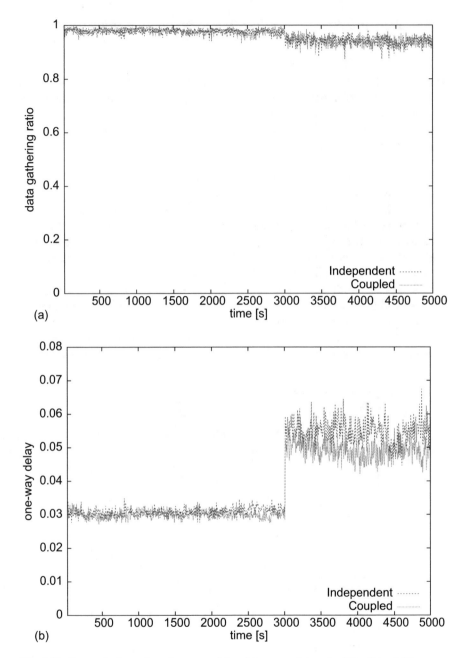

Fig. 3.11 Data gathering ratio and one-way delay: (**a**) average data gathering ratio and (**b**) average one-way delay

3.4.3 Network Resource Allocation to Multiple Applications on Multiple Vehicles

The first example of layered multipath routing involves combining the same control methods operating on different layers, while the second example of routing-aware clustering involves combining different control methods operating on different layers. The third example introduced here involves combining the same control methods operating on a single entity but competing for shared and limited network resources.

Specifically, we consider a vehicular network in which various applications, such as WWW, Voice over Internet Protocol (VoIP), and video streaming, operate on each vehicle, as illustrated in Fig. 3.12. Each of the applications has different requirements for the quality of communication, that is, Quality of Service (QoS), in terms of the bandwidth, delay jitter, and transmission cost. Here, we assume that dedicated short-range communications (DSRC), Wi-Fi, Worldwide Interoperability for Microwave Access (WiMAX), and cellular networks are available for use in the area. A Wi-Fi network is only available around an intersection of roads, whereas the other networks are available in the entire region. To maximize performance, applications compete for shared network resources. Furthermore, there exists competition among vehicles in the region. Therefore, the problem is to allocate heterogeneous network resources to vehicles in the region and to heterogeneous applications on

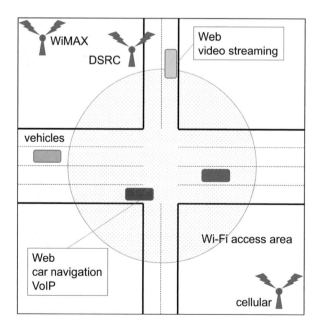

Fig. 3.12 Wireless network allocation to vehicles and applications

each vehicle. This must be accomplished by a distributed mechanism because no central server can manage such a dynamic and complex situation.

For adaptive inter-application resource allocation in a vehicle, we use a coupling model $\frac{d\mathbf{x}_i}{dt} = f_i(\mathbf{x}_i)\alpha + \eta_i$ (Eq. (3.7)). This model signifies that applications share a common activity represented as α, which is defined as a function of the effectiveness of resource allocation to each application on a vehicle. Adaptive and efficient resource allocation to vehicles is accomplished by the Yuragi-based adaptive behavior of vehicles as in Yuragi-based routing, where the performance of the entire system is maximized by the autonomous and distributed behavior of each session to maximize its activity [4].

We assume that each vehicle has a controller that mediates the resource allocation of applications running on the vehicle. At regular intervals, the applications declare their QoS requirements in terms of the required bandwidth, tolerable delay jitter, and acceptable transmission cost. The controller obtains information about the current status of available networks, such as the available bandwidth, delay jitter, and transmission cost, using cognitive radio technology. Next, the controller evaluates the degree of satisfaction Q_i of each application i with the allocated resources. Q_i is derived from the weighted average and weighted variance of the degree of satisfaction of each QoS requirement. Then, the degree of satisfaction S of a vehicle is derived as

$$S = \frac{\sum\limits_{i=1}^{N} \omega_i Q_i}{1 + b\sigma_Q},$$

where N is the number of applications running on a vehicle, and σ_Q is the weighted standard deviation defined as

$$\sigma_Q = \sqrt{\sum_{i=1}^{N} \omega_i \left(\bar{Q} - Q_i\right)^2},$$

where $\bar{Q}$ is the average of Q_is. Weight ω_i $(0 < \omega_i < 1)$ reflects the importance of application i in the vehicle. For example, while a user talks over a VoIP application, Web and video streaming have lower priority in resource allocation.

Using the degree of satisfaction S of a vehicle, the controller derives the activity α as

$$\alpha = \frac{\sum\limits_{k=0}^{W-1} \alpha_{h-k}^*}{W}.$$

Here, α_h^* is called the instant activity derived at the hth control timing, and W is the measurement window. Therefore, activity α is the average of the instant activity over the past W measurements. The instant activity α_h^* of the current measurement is derived from the degree of satisfaction S of a vehicle using a hysteresis function of the play model [2]:

$$\alpha_h^* = \frac{1}{P} \sum_{l=1}^{P} h_l(p_l(S')).$$

Here, $S' = 1.5S - 0.5$ to have an effective operating range by shifting the hysteresis curve to the right. P is the number of play hysterons, and p_l ($1 \le l \le P$) is a play hysteron operator defined as

$$p_l(S) = \max\left(\min\left(p_l^-, S' + \varsigma\right), S - \varsigma\right),$$

where p_l^- is the value of p_l at a previous time point, and ς is the width of the play hysteron. h_l is the shape function of the play hysteron p_l, which is defined as a sigmoid function by the following equation:

$$h_l(p_l(S)) = \frac{1}{1 + \exp(-gp_l(S))}.$$

Here, g ($g > 0$) is the gain of the sigmoid function.

A hysteresis function is used to prevent control from being overly sensitive to sudden small changes in the degree of satisfaction. In the application examples provided above, smoothing measurements or activity plays a similar role in avoiding sensitive control. Figure 3.13 presents an example of a trajectory taken by the instant

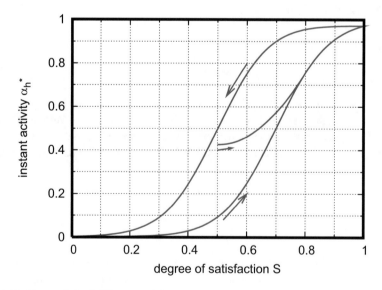

Fig. 3.13 Hysteresis loop of instant activity

activity, which draws a loop. Starting from any point, the trajectory is entrained to the loop. When the degree of satisfaction of a vehicle increases from zero, the instant activity remains low to allow the controller to seek better resource allocation until the degree of satisfaction becomes sufficiently large. Then, the instant activity exponentially increases beyond a certain point to accelerate convergence to a desirable attractor. However, when the environmental conditions change and the current satisfactory allocation is no longer appropriate, the degree of satisfaction decreases. However, the instant activity remains high until a certain point. Therefore, sudden small changes do not trigger exploration of better resource allocation, and control stability is maintained in a dynamic environment. However, once the degree of satisfaction is lower than a certain point, the instant activity drastically decreases to initiate a random walk search to adapt to the environmental changes.

Based on the activity α, the state vectors are updated. A state vector $\mathbf{m}_i$ of application i consists of

$$\mathbf{m}_i = \left(m_{i,1}, \ldots, m_{i,M} \right),$$

where M is the number of wireless networks available to a vehicle. Here, $\mathbf{m}_i$ is updated by Eq. (3.2), replacing m_i with $m_{i,j}$. Then, the wireless network of index j whose state value $m_{i,j}$ is the maximum in $\mathbf{m}_i$ is selected and assigned to application i.

Simulation results are presented in Fig. 3.14 for a video application running on a vehicle. We considered an area of 300×300 m^2 that had a horizontal road with

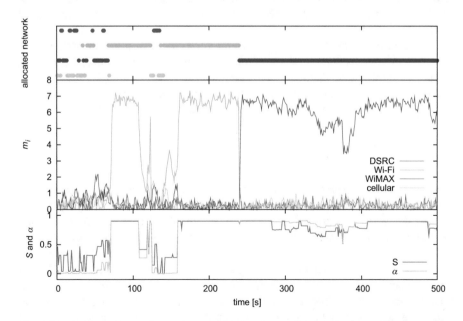

Fig. 3.14 Time variation of allocated network, state values, activity, and degree of satisfaction

four lanes and a vertical road with two lanes crossing at the center of the area (see Fig. 3.12). A total of 60 vehicles drove at a speed of 40 km/h on the horizontal road and 20 km/h on the vertical road and were affected by traffic lights at the intersection. The area was vertically and horizontally connected, that is, a torus. The Wi-Fi network had a limited access area that was circular with a radius of 100 m and centered jitter. In addition, the network transmission costs were set to (4 Mbps, 100 ms, 10^{-7} unit/b), (20 Mbps, 500 ms, 10^{-9} unit/b), (40 Mbps, 200 ms, 10^{-8} unit/b), and (2 Mbps, 100 ms, 10^{-5} unit/b) on DSRC, Wi-Fi, WiMAX, and cellular networks, respectively. Three applications ran on each vehicle, whose QoS requirements for the bandwidth, delay jitter, and transmission cost were set to (300 kbps, 10,000 ms, 0.1 unit/s), (64 kbps, 150 ms, 1 unit/s), and (3000 kbps, 1000 ms, 0.1 unit/s) for Web, VoIP, and video applications, respectively. These applications had different QoS weights as follows: (0.3, 0.1, 0.6), (0.5, 0.4, 0.1), and (0.6, 0.1, 0.3), respectively. For example, the Web application valued low-cost networking, while the VoIP application valued guaranteed bandwidth and small jitter.

Initially, the activity was low (bottom of Fig. 3.14), and as a result, the state values randomly changed (center of Fig. 3.14). Because the wireless network with the largest state value was allocated to the video application, we observed frequent changes in the allocated network (top of Fig. 3.14). However, as time passed, allocation converged to the Wi-Fi network, which offered the second-largest bandwidth with the lowest cost. After some time, as the vehicle exited the Wi-Fi access area, the activity gradually decreased, and other networks were tentatively allocated. The video application resumed using the Wi-Fi network once; however, it finally switched to the WiMAX network, which provided a stable, high-speed connection at a lost cost.

Next, we compared Yuragi-based resource allocation with an ad hoc optimization method. At regular intervals, a controller with the ad hoc optimization method obtained information about applications and networks, as in the Yuragi-based method. Then, it derived and adopted the optimal resource allocation with which the degree of satisfaction of a vehicle was maximized by solving an optimization problem. That is, the vehicles used resources in a greedy and selfish manner. Figure 3.15 presents the mean and variance of the degree of satisfaction of vehicles against a different number of vehicles from 10 to 120. Lines correspond to the average degree of satisfaction of vehicles on the left y-axis, while crosses correspond to the average variance of the degree of satisfaction of vehicles on the right y-axis as an indicator of the fairness of resource allocation among vehicles.

As illustrated in Fig. 3.15, the average degree of satisfaction with the ad hoc optimization method suddenly decreased when the number of vehicles exceeded 50. This is because the network resources were insufficient to accommodate too many vehicles with an equally high degree of satisfaction. In contrast, the Yuragi-based resource allocation method sustainably provided network service to a large number of vehicles. This is because the Yuragi-based method did not insist on determining the optimal solution. Due to its stochastic behavior, the Yuragi-based method sometimes converged to the near-optimal solution, causing an increase

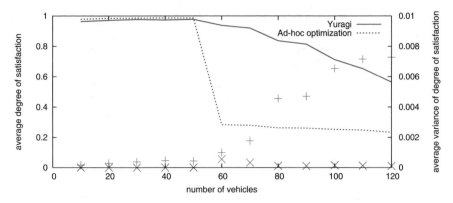

Fig. 3.15 Average degree of satisfaction and average variance of degree of satisfaction

in the average variance of the degree of satisfaction. The QoS requirements of all applications of all vehicles were not fully satisfied; however, they achieved moderately high performance.

3.5 Exploration of Better Attractors

As reviewed in the above subsections, Yuragi-based network control methods are adaptable to a dynamic environment, robust to failures, sustainable in infeasible conditions, and stable. Such characteristics are obtained by reinforcement learning with function $f(\mathbf{x})$ incorporated with random walk search with a noise term η. Activity α in the combined equation $\frac{d\mathbf{x}}{dt} = f(\mathbf{x})\alpha + \eta$ balances reinforcement learning and random walk search by providing a bias to either function $f(\mathbf{x})$ or noise term η depending on the quality of the current state $\mathbf{x}$ with respect to the current environmental conditions.

However, the structure of the Yuragi model is inherently problematic. Once the system converges to an attractor, it does not change its state until the activity decreases sufficiently. Therefore, even if another attractor has better performance, the system is not aware of it and cannot move to it. For example, assume that a session considers one path as primary and uses it for packet transmission with Yuragi-based multipath routing. In this case, the activity is high. Then, for a certain reason, such as a flash crowd, the primary path becomes heavily congested. Driven by the decreased activity, the session randomly tries other alternative paths. The session then finds a new optimal path and begins to use it as a primary path through reinforcement by the increased activity. However, at that time, the former primary path becomes vacant due to the disappearance of sudden traffic congestion. The former path can provide the session with better performance than the current primary path; however, the session is not aware of this due to the stability of the current attractor with high activity. For the system to find a new optimal attractor, it must examine alternative attractors and evaluate the activity.

Another problem inherent to the Yuragi model is that increasing the intensity of noise to allow a stable system to search for a new optimal attractor does not necessarily improve performance. Instead, it often degrades the control stability, and the performance fluctuates because the system continuously moves from one attractor to another. It is thus necessary for the system to remain at an attractor for some time. Changing an attractor affects other systems competing for shared resources and causes a cascading reaction, and measurements taken during such a transient phase are unreliable. Therefore, to obtain a reliable observation of the system conditions and derive the activity, it is necessary to wait for the entire system to stabilize. In addition, there is a possibility that the system will become trapped at a local optimum. During the random search phase, the performance is likely to be low. When the system finds an attractor that provides better performance than the visited attractors, the activity increases. Consequently, the system converges to the found attractor even if the obtained performance is lower than that of the former stable attractor.

To address this problem, we consider a Yuragi model with stochastic attractor exploration with short-term memory [12]. With the new extended model, even when the activity is high, the system temporarily tries another attractor, which we call exploration, by increasing the noise intensity. However, a mere increase in noise disturbs convergence to a desirable attractor. Therefore, we adaptively adjust the intensity of noise to explore attractors with moderate frequency and guarantee convergence. Furthermore, for the system to return to a former desirable attractor from a temporary poorer attractor that it reaches, we introduce short-term memory into the model.

The extended new Yuragi model is given as

$$\frac{dm_i}{dt} = \frac{\alpha(\beta\alpha^\gamma + \phi^*)}{1 + \max\limits_{1 \le j \le M} m_j^2 - m_i^2} - \alpha m_i + \sigma \eta_i - K(1-\alpha)(m_i - x_i)R^C. \quad (3.14)$$

Here, two new factors are added to Eq. (3.2). First, noise strength σ is introduced to strike a balance between active exploration and stabilization of a system. It is derived by the following equations:

$$\begin{cases} \sigma = \sigma_{min}, & \text{when path change occurs} \\ \frac{d\sigma}{dt} = -\frac{1}{\tau_\sigma}(\sigma - \sigma_{min} + \delta)(\sigma - \sigma_{max}), & \text{otherwise} \end{cases},$$

where σ_{min} and σ_{max} are constant parameters, and $0 < \sigma_{min} < \sigma_{max}$. When a session changes a path to use, the noise strength σ is set to its minimum, σ_{min}, to allow the session to remain at the path until the noise strength σ becomes sufficiently large. This is intended to avoid frequent path changes and promote reliable measurement. The noise strength σ then gradually increases σ_{min} to σ_{max} with time constant τ_σ. δ is a small constant to prevent the noise strength σ from being fixed to σ_{min}.

The second factor added to Eq. (3.2) is the short-term memory expressed by the fourth term on the right-hand side of Eq. (3.14). K and R are constants related to memory strength, and $\mathbf{x} = (x_1, \ldots, x_M)$ is the memory of state vector $\mathbf{m}$. The vector $\mathbf{x}$ develops as

$$\frac{dx_i}{dt} = -\frac{1}{\tau_x}(x_i - m_i) , \tag{3.15}$$

where τ_x ($\tau_x > 1$) is a time constant. The equation represents the exponential dependency of x_i on m_i, similar to the forgetting curve of human short-term memory [10]. When activity α is large at the current attractor during exploration, the memory term is less effective, and the system can remain at the attractor. However, when the activity is low ($\alpha < 1$), the memory term returns the system state to the memorized state, whereas old memory expressed by the vector $\mathbf{x}$ gradually disappears by Eq. (3.15). R corresponds to the distance between the current state $\mathbf{m}$ and the state $\mathbf{x}$ in the memory as

$$R = \sqrt{\sum_{i=1}^{M}(m_i - x_i)^2} .$$

The influence of R becomes significant when the current effective attractor is no longer appropriate. Because the current state $\mathbf{m}$ is very similar to the memorized state $\mathbf{x}$, R is small, and the memory term becomes negligible, which allows the system to start a random walk.

Finally, the activity is derived as

$$\alpha(t) = \min\left(\left(\frac{y}{d(t)} \right)^n, 1 \right) ,$$

where

$$\frac{dy}{dt} = -\frac{1}{\tau_y}(y - d(t)) .$$

Here, $d(t)$ is the performance (e.g., delay) obtained at the tth control timing, y is the memory of the performance $d(t)$, n ($n \geq 1$) is an exponent parameter, and τ_y ($\tau_y > 1$) is a time constant.

Figure 3.16 presents an example of Yuragi model behavior with stochastic attractor exploration with short-term memory. At the beginning, State 2 is selected due to the high state value m_2, and memory x_2 changes in accordance with state value m_2. At approximately 6000 s, state value m_1 becomes larger than m_2, and the system examines State 1 due to active exploration caused by the increased σ (see Fig. 3.16b). Immediately, σ returns to σ_{min}. Because the performance of State 1 is low, the system soon returns to State 2 by being pulled back by the memory, which

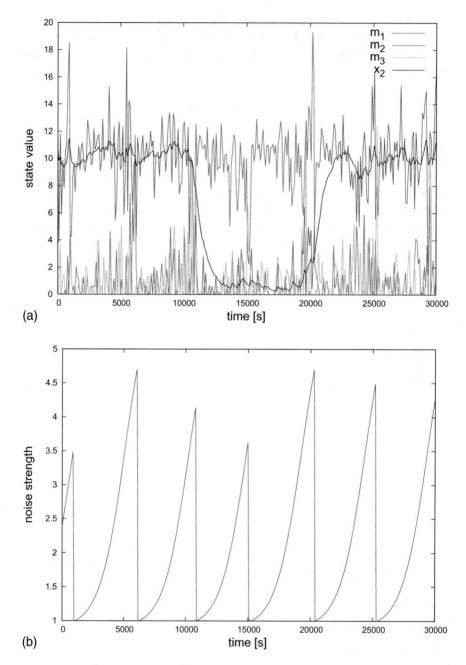

Fig. 3.16 Behavior of Yuragi model with short-term memory: (**a**) state values **m** and memory value x_2 and (**b**) noise strength σ

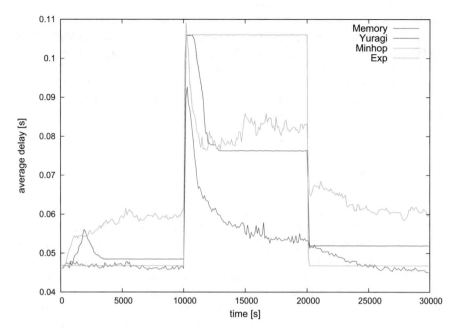

Fig. 3.17 Average delay

does not occur in the original Yuragi model. The activity decreases at approximately 10,000 s; consequently, the system changes to State 1. However, memory value x_2 remains high for some time to enable the system to return to State 2 if State 1 is found to be inappropriate. Memory value x_2 gradually decreases by a forgetting function because State 1 is satisfactory in this example.

Figure 3.17 presents the comparative results between a multipath routing method adopting the Yuragi model with stochastic attractor exploration with short-term memory and other methods. The network consisted of 128 nodes, and the capacity and propagation delay of each link were set to 10 Mbps and 10 ms, respectively. All pairs of nodes generated packets at a rate of 20 kbps. From 10,000 to 20,000 s, the amount of traffic between a pair of neighbor nodes was increased to increase the link utilization to 100%. The number of candidate paths was three (i.e., $M = 3$), and the control interval was 100 s. One-way delay was used as the indicator $d(t)$ of the quality of a path. The other parameters were set as follows: $\beta = 10$, $\gamma = 3$, $\sigma_{min} = 2$, $\sigma_{max} = 6$, $K = 0.4$, $C = 1$, $n = 2$, $\delta = 0.1$, $\tau_x = 8$, $\tau_y = 8$, and $\tau_\sigma = 50$. In Fig. 3.17, *Memory* refers to the new method, *Yuragi* refers to the original method described in Sect. 3.3, *Minhop* refers to a method that always stays on the shortest path, and *Exp* refers to a method that uses Eq. (3.14) but without the fourth memory term.

As illustrated in Fig. 3.17, the Memory method achieved the best performance. In contrast, Yuragi could adaptively move to a better path but failed to find a significantly better path, which Memory could accomplish, especially during the highly congested period. The flat delay of the Yuragi method signifies that the method remained at the suboptimal solution without further exploration. After 20,000 s, the Memory method successfully returned to the original path that was used before congestion. The reason why the delay of Memory slightly fluctuated was active exploration. However, short-term memory successfully pulled the session back to the former desirable path. In contrast, the Exp method continued searching for a better path and did not converge.

3.6 Conclusion

The Yuragi model combines reinforcement learning, which contributes to stability, and a noise-induced random walk, which searches for superior solutions. As discussed in this chapter, by applying the Yuragi model, highly robust, adaptive, and error-tolerant network control can be achieved without prior knowledge of a dynamic environment and without deterministic adaptation rules, similarly to biological systems. In essence, it is only necessary to define attractors corresponding to control options and the activity as an objective function of the control. It should be noted that the system must be tolerant of suboptimal solutions and occasional fluctuations when adopting Yuragi-based control. However, optimal performance cannot be maintained by any control mechanism in a complex, dynamic, and unpredictable environment. Yuragi-based control can maintain suboptimal performance, and the application area is not limited to network control.

References

1. Asvarujanon, N., Leibnitz, K., Wakamiya, N., Murata, M.: Robust and adaptive mobile ad hoc routing with attractor selection. In: Proceedings of 4th International Workshop on Adaptive and Dependable Mobile Ubiquitous Systems (ADAMUS 2010) (2010)
2. Bobbio, S., Milano, G., Serpico, C., Visone, C.: Models of magnetic hysteresis based on play and stop hysterons. IEEE Trans. Magn. **33**(6), 4417–4426 (1997)
3. Iwai, T., Wakamiya, N., Murata, M.: Error-tolerant and energy-efficient coverage control based on biological attractor selection model in wireless sensor networks. Int. J. Distrib. Sens. Netw. **8**(2), 971014 (2012)
4. Kajioka, S., Wakamiya, N., Murata, M.: Autonomous and adaptive resource allocation among multiple nodes and multiple applications in heterogeneous wireless networks. J. Comput. Syst. Sci. **78**, 1673–1685 (2012)
5. Leibnitz, K., Wakamiya, N., Murata, M.: Symbiotic multi-path routing with attractor selection. In: Proceedings of Workshop on Stochasticity in Distributed Systems (StoDis 2005) (2005)
6. Leibnitz, K., Wakamiya, N., Murata, M.: Resilient multi-path routing based on a biological attractor-selection scheme. In: Proceedings of the Second International Workshop on Biologically Inspired Approaches to Advanced Information Technology (Bio-ADIT 2006), pp. 48–63 (2006)
7. Leibnitz, K., Wakamiya, N., Murata, M.: Self-adaptive ad-hoc/sensor network routing with attractor-selection. In: Proceedings of IEEE GLOBECOM 2006 (2006)

8. Leibnitz, K., Wakamiya, N., Murata, M.: A bio-inspired robust routing protocol for mobile ad hoc networks. In: Proceedings of 16th International Conference on Computer Communications and Networks (ICCCN 2007), pp. 321–326 (2007)
9. Leibnitz, K., Wakamiya, N., Murata, M.: Biologically inspired self-adaptive multi-path routing in overlay networks. Commun. ACM **49**(3), 63–67 (2016)
10. London, I.D.: An ideal equation derived for a class of forgetting curves. Psychol. Rev. **57**(5), 295–302 (1950)
11. Mori, H., Wakamiya, N.: Bio-inspired cluster-based routing for wireless sensor networks. In: Proceedings of International Symposium on Nonlinear Theory and Its Applications (NOLTA2014), pp. 791–794 (2014)
12. Nakao, T., Teramae, J., Wakamiya, N.: An adaptive routing protocol with balanced stochastic route exploration and stabilization based on short-term memory. IEICE Trans. Commun. **E99-B**(11), 2280–2288 (2016)
13. Onzuka, N., Wakamiya, N., Murata, M.: Robust and lightweight routing with attractor selection. In: Proceedings of the 4th World Conference on Information Technology 2013 (2013)
14. Perkins, C.E., Belding-Royer, E.M., Das, S.R.: Ad hoc on-demand distance vector (AODV) routing, RFC3561 (2003)
15. Takeshita, E., Wakamiya, N.: Proposal and evaluation of attractor selection-based adaptive routing in layered networks. In: Proceedings of International Symposium on Ubiquitous Intelligence and Autonomic Systems (UIAS-2013) (2013)
16. Takeshita, E., Wakamiya, N.: Adaptive multipath routing for large-scale layered networks. In: Proceedings of the 17th Asia-Pacific Network Operations and Management Symposium (APNOMS 2015), pp. 221–226 (2015)
17. Waxman, B.M.: Routing of multipoint connections. IEEE J. Sel. Areas Commun. **6**(9), 1617–1622 (1988)
18. Yang, H., Wakamiya, N., Murata, M., Iwai, T., Yamano, S.: Autonomous and distributed mobility management scheme in mobile core networks. Wireless Netw. (2016). https://doi.org/10.4108/eai.3-12-2015.2262427

Chapter 4
Yuragi-Based Virtual Network Control

Yuki Koizumi

Abstract Reconfiguring a virtual network, which consists of a set of virtual links and routers, on top of a physical network is a promising approach to accommodate time-varying traffic. However, optimization-based approaches are generally incapable of quickly adapting to time-varying traffic due to their computational complexity. For swift adaptation to changes in traffic patterns and network failures, this chapter presents *Yuragi-based virtual network control*. This control is based on *attractor selection*, which models behavior whereby biological systems adapt to unknown changes in their surrounding environments. A biological system driven by attractor selection adapts to environmental changes, selecting *attractors*, in which the condition of the system is well suited for a certain environment, by using noise, also referred to as *Yuragi*. There are two main challenges in achieving virtual network control based on attractor selection. The first involves determining how to map the behavior of a biological system to virtual network control, while the second involves designing attractors for robust virtual network control. This chapter summarizes previous studies that solved these two research problems.

4.1 Introduction

The emergence of new application layer services, such as video on demand and user-generated content delivery, results in fluctuations in network environments, such as network failures and traffic changes, as well as increased Internet traffic. The Internet plays an increasingly vital role as social infrastructure; therefore, the ability to withstand and recover from various changes in network environments has become a crucial requirement.

Y. Koizumi (✉)
Graduate School of Information Science and Technology, Osaka University, Suita, Osaka, Japan
e-mail: ykoizumi@ist.osaka-u.ac.jp

One promising approach to accommodate fluctuation in network environments is to construct a *virtual network*, which consists of a set of virtual links and routers, on top of a physical network according to the current network environment. Several network virtualization technologies, such as overlay, multi-protocol label switching (MPLS), and generalized MPLS (GMPLS) networks, have been proposed [5]. Among such network virtualization technologies, this chapter focuses on wavelength-division multiplexing (WDM) networks, which offer a flexible virtual network infrastructure by using wavelength routing. WDM networks are suitable for accommodating high Internet traffic owing to their high link capacity; therefore, much research has been devoted to developing methods for carrying Internet traffic over wavelength-routed WDM networks [1, 6, 10, 15, 18, 19, 23, 29, 30]. In a wavelength-routed WDM network, optical transport channels, referred to as lightpaths, are established between routers via optical cross-connects (OXCs), and a set of lightpaths and routers forms a virtual network. Virtual network control, which configures a virtual network on the basis of a given network environment, has been investigated in several studies [21, 24].

By reconfiguring virtual networks, wavelength-routed WDM networks offer a means of adapting to changing network environments. Existing virtual network control methods, which are based on the control paradigm developed in the area of engineering, primarily consider a particular set of scenarios of environmental changes and use an algorithm to perform predefined countermeasures to those changes. Although these methods guarantee optimal performance for their assumed environments, they cannot achieve expected performance in the case of unexpected environmental changes.

A remarkable example of adapting to various environmental changes can be observed in a biological system, which is studied in the field of life sciences [13]. This chapter focuses on *attractor selection*, which models the behavior of organisms when they adapt to unknown changes in their surrounding environments and recover their conditions. Kashiwagi et al. [14] proposed an attractor selection model for an Escherichia coli cell that adapts to changes in nutrient availability. Furusawa and Kaneko [8] introduced another attractor selection model for explaining the adaptability of a cell consisting of a gene regulatory network and a metabolic network. The fundamental principle of attractor selection is that a system adapts to environmental changes by selecting an *attractor* suitable for the current surrounding environment. This selection mechanism in attractor selection is based on deterministic behavior and stochastic behavior that are controlled by simple feedback on the system condition.

The selection mechanism in attractor selection is one of the main differences between attractor selection and heuristic or optimization approaches in the field of engineering. In contrast to engineering systems, biological systems do not rely on predefined algorithms. Instead, they exploit stochastic behavior, which is environmental noise from various sources, to adapt to environmental changes. This chapter refers to attractor selection as *Yuragi-based control*, where *Yuragi* is the Japanese term for noise. On the one hand, biological systems do not guarantee optimal performance, whereas engineering systems generally achieve optimal performance for assumed environments. On the other hand, biological systems are capable of adapting to unknown environmental changes thanks to their stochastic behavior, whereas engineering systems cannot handle environmental changes that are not accounted for by their predefined algorithms. We therefore adopt attractor selection as the key mechanism in our virtual network control method to achieve adaptability to various environmental changes.

There are two main challenges in achieving virtual network control based on attractor selection. The first challenge involves determining how to map attractor selection in a biological system to virtual network control in a wavelength-routed WDM network. The second challenge involves determining how to design attractors for adaptive virtual network control. Previous studies [16, 17] solved the first research problem. Focusing on the similarity between a wavelength-routed WDM network, which consists of a physical and a virtual network, and a cell, which consists of a gene regulatory and a metabolic reaction network, we developed a virtual network control method based on attractor selection found in a cell [8]. In addition, we incorporated the knowledge of the Hopfield neural network [12] into the proposed method to embed attractors into the deterministic behavior of the proposed virtual network control method, focusing on the similarity between attractor selection in a cell and the Hopfield neural network. Another study [22] solved the second research problem by proposing a method for designing diverse attractors to be stored in the attractor structure to handle various environmental changes. This chapter summarizes this series of studies.

The remainder of this chapter is organized as follows. Section 4.2 introduces the concept of attractor selection, while Sect. 4.3 proposes an adaptive virtual network control method based on attractor selection. Section 4.4 presents an algorithm for designing attractors for adaptive virtual network control, while Sect. 4.5 summarizes related approaches. Finally, Sect. 4.6 concludes the chapter.

4.2   Attractor Selection

This section describes attractor selection, which is the core mechanism in our adaptive virtual network control method. The original model for attractor selection was introduced in [8].

4.2.1   Concept of Attractor Selection

A dynamic system driven by attractor selection uses noise to adapt to environmental changes. In attractor selection, attractors are a part of the equilibrium points in the phase space in which the system condition is preferable. The underlying mechanism consists of deterministic and stochastic behaviors. When the current system condition is suitable for the environment, i.e., the system state is close to one of the attractors, the deterministic behavior drives the system to the attractor. When the current system condition is poor, however, the stochastic behavior dominates over the deterministic behavior. While the stochastic behavior is dominant in controlling the system, the system state fluctuates randomly due to noise, and the system searches for a new attractor. When the system condition recovers and the system state approaches an attractor, the deterministic behavior again controls the system. These two behaviors are controlled by simple feedback on the system condition. Therefore, attractor selection adapts to environmental changes by selecting attractors using the stochastic behavior, the deterministic behavior, and the simple feedback. In Sect. 4.2.2, we introduce attractor selection that models the behavior of the gene regulatory and the metabolic reaction networks of a cell.

4.2.2   Cell Model

Figure 4.1 presents a schematic diagram of the cell model used in [8]. It consists of two networks: the gene regulatory network in the dotted rectangle at the top of the figure and the metabolic reaction network in the dotted rectangle at the bottom of the figure.

Each gene in the gene regulatory network has protein expression levels, and deterministic and stochastic behaviors in each gene control the expression level. The deterministic behavior controls the expression level due to the effects of activation and inhibition from the other genes. In Fig. 4.1, the effects of activation and inhibi-

Fig. 4.1 Schematic diagram of a gene regulatory and a metabolic reaction network

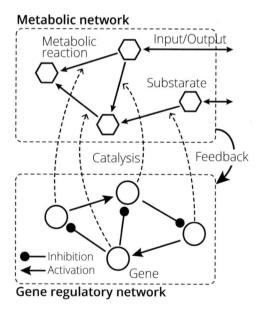

tion are indicated by triangular-headed and circular-headed arrows, respectively. In the stochastic behavior, inherent noise randomly changes the expression levels.

In the metabolic reaction network, metabolic reactions consume various substrates and produce new substrates. Proteins in the corresponding genes catalyze these metabolic reactions. In Fig. 4.1, metabolic reactions are illustrated as fluxes of substrates, and catalyses of proteins are indicated by dashed arrows. Changes in the concentrations of metabolic substrates are caused by metabolic reactions and the transportation of substrates from outside the cell. Some nutrient substrates are supplied from the environment by diffusion through the cell membrane.

The dynamics of the metabolic reactions determines the growth rate. Some metabolic substrates are necessary for cellular growth; thus, the growth rate is determined as an increasing function of their concentrations. The gene regulatory network uses the growth rate as feedback on the condition of the metabolic reaction network and controls the deterministic and stochastic behaviors using the growth rate. If the metabolic reaction network is in poor condition and the growth rate is low, the stochastic behavior dominates over the deterministic behavior, triggering a search for a new attractor. During this phase, the expression levels are randomly changed by noise, and the gene regulatory network searches for a state that is suitable for the current environment. After the condition of the metabolic reaction network is recovered and the growth rate increases, the deterministic behavior again drives the gene regulatory network to a stable state. Section 4.2.3 describes the mathematical model of attractor selection in greater detail.

4.2.3   Mathematical Model of Attractor Selection

The internal state of a cell is represented by a set of protein expression levels of n genes $(x_1, x_2, \ldots, x_n)$, and concentrations of m metabolic substrates $(y_1, y_2, \ldots, y_m)$. The dynamics of the protein expression level of the ith gene, x_i, is described as

$$\frac{dx_i}{dt} = \alpha \left(\varsigma \left(\sum_j W_{ij} x_j - \theta \right) - x_i \right) + \eta. \tag{4.1}$$

The first and second terms on the right-hand side represent the deterministic behavior of gene i, while the third term represents the stochastic behavior. In the first term, the regulation of the protein expression level of gene i by other genes is indicated by a regulatory matrix W_{ij}, which takes 1, 0, or -1 corresponding to activation, no regulatory interaction, or inhibition of gene i by gene j, respectively. The rate of increase in the expression level is given by the sigmoidal regulation function, $\varsigma(z) = 1/(1 + e^{-\mu z})$, where $z = \sum W_{ij} x_j - \theta$ is the total regulatory input with threshold θ for increasing x_i, and μ denotes the gain parameter of the sigmoid function. The second term represents the rate of decrease in the expression level of gene i. This term signifies that the expression level decreases depending on the current expression level. The last term on the right-hand side of Eq. (4.1), η, represents molecular fluctuation, which is Gaussian white noise. Noise η is independent of the production and consumption terms, and its amplitude is constant. The change in expression level x_i is determined by the deterministic behavior, i.e., the first and second terms in Eq. (4.1), and the stochastic behavior η. The deterministic and stochastic behaviors are controlled by the growth rate α, which represents the condition of the metabolic reaction network.

In the metabolic reaction network, metabolic reactions, which are an internal influence, and the transportation of substrates from the outside of the cell, which is an external influence, determine the changes in the concentrations of metabolic substrates y_i. Proteins of the corresponding genes catalyze the metabolic reactions, and the expression level x_i determines the strength of the catalysis. A high x_i accelerates the metabolic reaction, while a low x_i suppresses it. In other words, the gene regulatory network controls the metabolic reaction network through catalysis.

Some of the metabolic substrates are necessary for cellular growth. The growth rate of the cell, α, is determined as an increasing function of the concentrations of these vital substrates. The gene regulatory network uses α as feedback on the condition on the metabolic reaction network and controls the deterministic and stochastic behaviors. The feedback value is referred to as *activity* in the framework of attractor selection. If the concentrations of the required substrates decrease due to changes in the concentrations of nutrient substrates outside the cell, α also decreases. By reducing α, the effects of the first and second terms in Eq. (4.1) on the dynamics of x_i decrease, and the effects of η increase. Thus, x_i fluctuates randomly, and the gene regulatory network searches for a new attractor. The fluctuation in x_i leads to changes in the rate of metabolic reactions via the protein catalysis. When the concentrations of the required substrates again increase, α also increases. Then, the first and second terms in Eq. (4.1) again dominate the dynamics of x_i over the stochastic behavior, and the system converges to the state of the attractor. Since we mainly use the gene regulatory network, we omit a detailed description of the metabolic reaction network from this chapter, and readers can refer to [8] for additional information. Section 4.3 describes the proposed virtual network control method based on the attractor selection model.

4.3   Virtual Network Control Based on Attractor Selection

In this section, we present a virtual network control method based on the attractor selection model of a gene regulatory network. We first introduce the network model used in this chapter and then describe our proposed method.

4.3.1 Virtual Network Control

A physical network consists of nodes having routes overlaying OXCs, with the nodes interconnected by optical fibers, as illustrated in Fig. 4.2a. Optical demultiplexers allow optical signals to be dropped at routers, and OXCs allow optical signals to pass through them. In such wavelength-routed WDM networks, nodes are connected with dedicated virtual circuits called lightpaths. Virtual network control configures lightpaths between routers via OXCs, and these lightpaths and routers form a virtual network, as illustrated in Fig. 4.2b. When lightpaths are configured in the WDM network, as illustrated in Fig. 4.2a, the virtual network in Fig. 4.2b is formed. The IP network uses a virtual network as its network infrastructure and transports IP traffic on the virtual network. By reconfiguring virtual networks, that is, by establishing lightpaths, wavelength-routed WDM networks offer the means to adapt to changing network environments. Determining how to reconfigure virtual networks is thus indispensable to develop an adaptive virtual network control method.

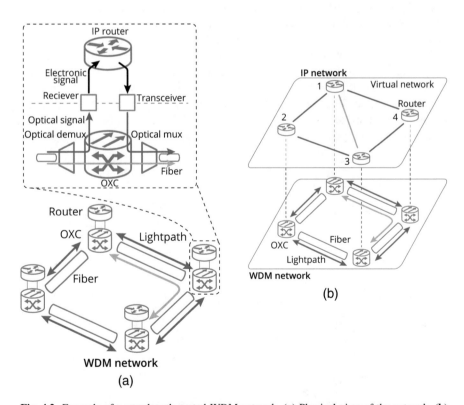

Fig. 4.2 Example of a wavelength-routed WDM network. (**a**) Physical view of the network. (**b**) Logical view of the network, where the lower layer represents a WDM network, which consists of OXCs and fibers, and the upper layer represents an IP network, which uses a virtual network constructed due to the wavelength-routing capability as network infrastructure

4.3.2 Overview of Virtual Network Control Based on Attractor Selection

In a cell, the gene regulatory network controls the metabolic reaction network, and the growth rate, which is the condition of the metabolic reaction network, is recovered when it is degraded due to changes in the environment. Similarly, the main objective of our virtual network control method is to recover the performance of the IP network by appropriately constructing virtual networks when the performance is degraded due to changes in the network environment. We therefore interpret the gene regulatory network as a WDM network and the metabolic reaction network as an IP network, as illustrated in Fig. 4.3. Using the stochastic behavior, our virtual network control method adapts to various changes in the network environment by selecting suitable attractors, which correspond to virtual networks in our method, for the current network environment, and the performance of the IP network can be recovered after it has degraded due to network failures.

A flowchart of the proposed virtual network control method is presented in Fig. 4.4. Our proposed approach works on the basis of periodic measurements of the IP network condition. One example of a condition measure is the load on links, which is the traffic volume on links. The link load is converted into activity, which is the value that controls the deterministic and stochastic behaviors. We describe the activity and the deterministic and stochastic behaviors in Sect. 4.3.3. The proposed method controls the deterministic and stochastic behaviors in the same way as attractor selection depending on the activity. The method constructs a new virtual network according to the system state of attractor selection, and the constructed virtual network is applied as the new infrastructure for the IP network. By sending

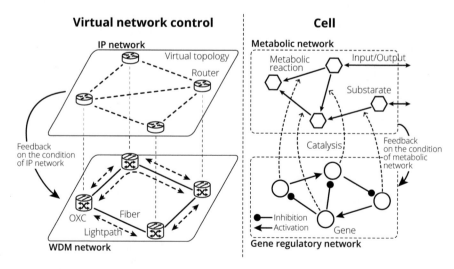

Fig. 4.3 Application of attractor selection to virtual network control

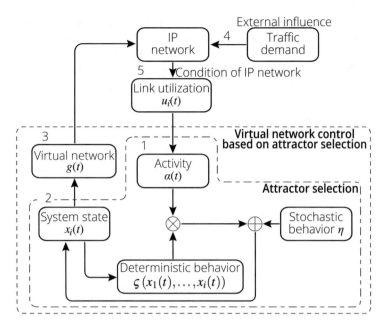

Fig. 4.4 Flowchart of virtual network control based on attractor selection

traffic over this new virtual network, the link load on the IP network is changed, and the method retrieves the link load to determine the IP network condition.

4.3.3 Dynamics of Virtual Network Control

We place genes on all candidates of possible lightpaths, where l_i denotes the ith lightpath. Each gene has its protein expression level x_i, and the l_i is controlled by x_i. The dynamics of x_i is described as

$$\frac{dx_i}{dt} = \alpha \left(\varsigma \left(\sum_j W_{ij} x_j \right) - x_i \right) + \eta, \qquad (4.2)$$

where η represents white Gaussian noise, $\varsigma(z) = 1/(1 + \exp(-z))$ is the sigmoidal regulation function, and the activity α, which is equivalent to the growth rate in cells introduced in Sect. 4.2, represents the IP network condition. We define α below. We use the same formula as Eq. (4.1) to calculate x_i, which is used to determine whether l_i is established. In our method, we establish l_i if $x_i > 0.5$; otherwise, we do not establish l_i. Therefore, our method interprets x_i as the virtual network.

In a cell, α represents the condition of the metabolic reaction network, and the gene regulatory network seeks to optimize α. Our method uses the maximum link utilization on the IP network as a metric representing the condition of the IP network. To obtain the maximum link utilization, we collect the traffic volume on all links and select the maximum value. This information is quickly and directly retrieved using simple network management protocol (SNMP). The activity must be an increasing function of the condition of the target system (in this case, the IP network), as mentioned in Sect. 4.2. Note that any metric that indicates the condition of an IP network, such as average end-to-end delay, average link utilization, and throughput, can be used to define α. We employ maximum link utilization as the IP network condition because it is a major performance metric for virtual network control that has been used in many studies [9, 24]. We convert the maximum link utilization on the IP network, $u_{\max}$, into α as follows:

$$\alpha = \frac{1}{1 + \exp\left(\delta\left(u_{\max} - \zeta\right)\right)}, \tag{4.3}$$

where δ represents the gradient of this function, and the constant ζ is the threshold for α. If the maximum link utilization is greater than ζ, α rapidly approaches 0 due to the unsatisfactory condition of the IP network. Therefore, the dynamics of our virtual network control method is governed by noise and the search for a new attractor. When the maximum link utilization is less than ζ, we rapidly increase α to improve the maximum link utilization.

A smooth transition between the current virtual network and the newly calculated virtual network is another important problem in virtual network control, as discussed in [7, 31]. Our virtual network control method constructs a new virtual network on the basis of the current virtual network, and the difference between these two virtual networks is given by Eq. (4.2). A high degree of activity signifies that the current system state x_i is near the attractor, which is one of the equilibrium points in Eq. (4.2); therefore, the difference given by this equation is close to zero. Consequently, our virtual network control method makes small changes to the virtual network, enabling adaptation to changes in the network. When there is a low degree of activity due to the poor IP network condition, the stochastic behavior dominates over the deterministic behavior. Thus, x_i fluctuates randomly due to noise η to search for a new virtual network that has higher activity, i.e., the lower maximum link utilization. Our method makes large changes to the virtual network to efficiently discover a suitable virtual network from a multitude of possible virtual networks. In this way, the proposed method modifies virtual networks depending on the maximum link utilization in the IP network and adapts to changes in traffic demand. We have already demonstrated that this method achieves smooth transition between virtual networks in [16].

The proposed method constructs virtual networks according to x_i, and x_i converges to one of attractors, which are a part of the equilibrium points in the phase space; therefore, the definition of attractors is a challenging and essential aspect of our proposed method. Section 4.3.4 presents how to define attractors in the phase space.

4.3.4   Attractor Structure

The regulatory matrix W_{ij} in Eq. (4.2) is an important parameter since it determines the locations of attractors in the phase space, referred to as the *attractor structure*. Our method selects one of the attractors according to Eq. (4.2) and constructs a virtual network corresponding to the selected attractor. Hence, defining W_{ij} is a challenging research problem. To define arbitrary attractors in the phase space, we use knowledge about the Hopfield neural network, which has a similar structure to gene regulatory networks.

The dynamics of our proposed method is expressed by Eq. (4.2). From the perspective of dynamical systems, α is regarded as a constant value that determines the convergence speed, and noise η is Gaussian white noise with mean 0. These values do not affect the equilibrium points, i.e., attractors in our method, in the phase space. Therefore, the equilibrium points are determined by the following differential equation:

$$\frac{d}{dt}x_i = \varsigma \left(\sum_j W_{ij} x_j \right) - x_i.$$

This is the same formula as for a continuous Hopfield neural network [12]. We therefore use methods to use Hopfield neural networks as content-associative memory to store arbitrary attractors in the phase space [3, 25].

Suppose that we store a set of virtual networks $g_k \in G$ in the phase space defined by Eq. (4.2). Let $x^{(k)} = (x_1^{(k)}, x_2^{(k)}, \ldots, x_i^{(k)})$ be the vector of the expression levels corresponding to virtual network g_k. To store $x^{(k)}$ in the phase space, we adopt the method introduced in [3], which stores patterns in the phase space by orthogonalizing them. Due to space limitations, we omit a detailed description of this method, and readers refer to [3] for a more detailed description. We store m virtual networks, $x^{(1)}, x^{(2)}, \ldots, x^{(m)}$, in the phase space. Let X be a matrix of which rows are $x^{(1)}, x^{(2)}, \ldots, x^{(m)}$. The regulatory matrix $W = \{W_{ij}\}$, of which attractors are $x^{(1)}, x^{(2)}, \ldots, x^{(m)}$, is defined as

$$W = X^+ X, \tag{4.4}$$

where X^+ is the pseudo-inverse matrix of X.

Although pattern orthogonalization results in the high stability of stored patterns [3], our method can also use more straightforward memorization approaches, such as Hebbian learning [11]. Using Hebbian learning, W_{ij} is defined as follows:

$$W_{ij} = \begin{cases} 0 & \text{if } i = j \\ \sum_{g_s \in G} (2x_i^{(s)} - 1)(2x_j^{(s)} - 1) & \text{otherwise.} \end{cases} \tag{4.5}$$

Our method can use either Eq. (4.4) or (4.5) to define the attractor structure. We mainly use Eq. (4.4) to define the regulatory matrix, as Baram [3] reported that pattern orthogonalization results in higher memory capacity than Hebbian learning. In Sect. 4.4, we describe a method for dynamically reconfiguring the attractor structure to adapt to a dynamic network environment.

4.4 Attractor Structure Design

According to Eq. (4.2), the expression levels converge to one of the attractors stored in the attractor structure defined in Eq. (4.4). Convergence signifies that the proposed method constructs any of the stored virtual networks. One approach to adapting to various changes in the network environment is to store all possible virtual networks as attractors. This approach is, however, impossible due to the memory capacity limitations of the Hopfield network [3, 11]. We therefore must select candidates for virtual networks embedded in an attractor structure.

There are two approaches to overcome the limitation on the number of attractors: The first is to reconfigure the attractor structure dynamically in accordance with the current environment, while the second is to design diverse attractors to be stored in the attractor structure so that they cover various environmental changes. This section presents methods to design virtual networks embedded in an attractor structure as attractors. Section 4.4.1 formulates the problem of designing an attractor structure, while Sect. 4.4.2 proposes a method to reconfigure an attractor structure dynamically. Sections 4.4.3 and 4.4.4 describe other methods for designing diverse attractors.

4.4.1 Problem Formulation

Suppose that we have a physical network consisting of n nodes. Under the assumption that lightpaths can be established between any node pairs, the number of possible lightpaths is approximately n^2. In this case, the number of possible topologies of virtual networks, which is the size of the solution space of virtual network control, is 2^{n^2}. In contrast, the number of attractors stored in an attractor structure, of which dimension is n^2, is approximately 10% of n^2 [3]. The problem is thus to select $0.1n^2$ virtual networks from 2^{n^2} possibilities so that the following objective is satisfied.

The objective of selecting virtual networks is to achieve adaptability to various environmental changes; therefore, the attractor structure must have diverse virtual networks as attractors. The problem of designing attractors thus involves selecting diverse $0.1n^2$ virtual networks so that any of them can accommodate a variety of environmental changes.

4.4.2 Dynamic Reconfiguration of Attractor Structure

It is crucial to develop a method for increasing the number of attractors suitable for the current network environment in the phase space. By incorporating new attractors suitable for the current network environment into the attractor structure, we achieve adaptability to various environmental changes when only a limited number of attractors are stored in the attractor structure.

To achieve this, we update the regulatory matrix W_{ij} in the case that our proposed method finds a virtual network that is suitable for the current network environment and is not a member of G when α is low. Before describing this approach in detail, we present the control flow of our proposed method in Fig. 4.4. For simplicity, we use the terms used in Eqs. (4.2), (4.3), and (4.4) with time t, e.g., the expression level at time t is expressed as $x_i(t)$. In Step 1 of Fig. 4.4, we calculate α from the maximum link utilization at time t, $\alpha(t)$. In Step 2, we update $x_i(t)$ with Eq. (4.2). Next, we convert $x_i(t)$ into the virtual network $g(t)$ and provide it as the network infrastructure of the IP network in Step 3. In Step 4, traffic flows on this virtual network $g(t)$. In Step 5, the resulting flow of traffic determines link utilization $u_i(t)$ on lightpath l_i. Then, we calculate α at time $t + 1$, $\alpha(t + 1)$, from $u_i(t)$. $\alpha(t + 1)$ indicates the quality of virtual network $g(t)$, which is converted from x_i. We can thus determine the quality of virtual network $g(t)$ for the current network environment by observing $\alpha(t + 1)$.

We can determine whether the performance on virtual network $g(t)$ is sufficiently high if $\alpha(t + 1) > A$, where A is a threshold value. We add $g(t)$ to the set of attractors G if $\alpha(t + 1) > A$, $\alpha(t) \leq A$, and $g(t) \notin G$. Then, we update W_{ij} with Eq. (4.4). We add another condition $\alpha(t) \leq A$ to prevent unnecessary updates of W_{ij}. The expression levels $x_i(t)$ always fluctuate due to the constant noise term, η, in Eq. (4.2). Even if α is high and the deterministic behavior dominates the dynamics of x_i over the stochastic behavior, x_i changes slightly; thus, our method sometimes constructs a slightly different virtual network from the already constructed network. Without the condition $\alpha(t) \leq A$, this behavior generates many attractors similar to one of the stored attractors, resulting in a lack of diversity of attractors. Therefore, the condition $\alpha(t) \leq A$ is necessary, and the algorithm adds $g(t)$ as a new attractor to the attractor structure only if the stochastic behavior dominates the dynamics of x_i.

Finally, we describe the deletion of attractors. Because the memory capacity of the Hopfield network is limited, as mentioned above, we cannot keep all of the virtual networks added in the attractor structure. We must delete some of the stored attractors when we add virtual network $g(t)$ as a new attractor. For this purpose, we use the first-in first-out (FIFO) policy for managing attractors when a new attractor is added Although we can use a more sophisticated policy to manage attractors, our method achieves sufficient adaptability with the FIFO policy, as discussed in [17].

4.4.3   Design of Diverse Attractor Structures

This section presents an algorithm for selecting diverse attractors. This algorithm is summarized as follows:

1. Generate virtual networks with the isomorphic graph structure of a given virtual network.
2. Classify the generated virtual networks into groups on the basis of their characteristics.
3. Select representative virtual networks from each group.

The principle behind this algorithm is that virtual networks that are derived from a desirable virtual network have the same desirable characteristics. In our case, desirable virtual network signifies that a virtual network accommodates a given traffic demand with low link utilization. Let us assume that a virtual network g is designed to accommodate a given traffic demand matrix T with a sophisticated algorithm, such as an optimization-based algorithm. In the first step, the algorithm derives isomorphic virtual networks from g assuming that the generated virtual networks accommodate T', which is different from T. Next, the algorithm generates clusters of virtual networks with similar characteristics. Finally, the algorithm selects a representative virtual network from each group and embeds it in the attractor structure as an attractor. The steps of the algorithm are described below.

Step 1. Generation of Isomorphic Virtual Networks   The algorithm first generates virtual networks that have an isomorphic graph structure of a given virtual network g. Such virtual networks are referred to as isomorphic virtual networks of g, hereinafter. Figure 4.5 presents an example of isomorphic virtual networks. Virtual network g_1 consists of five nodes $N_0, N_1, \ldots,$ and N_4, and g_2 and g_3 are isomorphic virtual networks of g_1. g_2 is generated by shifting N_0 of g_1 to N_1, N_1 to N_2, N_2 to N_3, N_3 to N_4, and N_4 to N_0. g_3 is also generated from g_1 in the same way. Virtual networks that do not satisfy the constraints on resources in the physical network, such as the number of ports on a router, are eliminated from the candidate pool. In this process, the algorithm generates at most $n!$ virtual networks from a given virtual network.

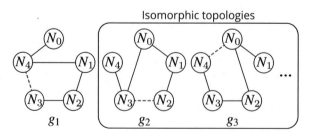

Fig. 4.5 Example of isomorphic virtual networks

Let us assume that a virtual network g_1 is designed to accommodate a given traffic demand matrix T_1 with a sophisticated algorithm, such as an optimization-based algorithm. The load on the link between N_3 and N_4, indicated by a dashed line in Fig. 4.5, is highest but is sufficiently low because g_1 is designed for T_1. Suppose that a traffic demand matrix T_2 is generated by randomly shuffling all the elements in T_1. In this case, at least one of the isomorphic virtual networks, such as g_2, is capable of accommodating T_2 because g_2 is derived from g_1 by shuffling the nodes in g_1. This implies that the isomorphic virtual networks of a well-designed virtual network are capable of accommodating changing traffic demand. In this study, the proposed algorithm uses I-MLTDA [2], which is a heuristic algorithm to design virtual networks for a given traffic demand matrix. However, the algorithm must further reduce the number of candidates for virtual networks because the number of the isomorphic virtual networks derived from g is still too high. Hereinafter, G denotes the set of isomorphic virtual networks, including g.

Step 2. Classify Virtual Networks To further reduce the number of virtual network candidates, the algorithm classifies the isomorphic virtual networks into groups on the basis of the characteristics of the networks and selects one network from each group as a representative virtual network. The algorithm uses edge betweenness centrality, which is the number of shortest paths that pass through a link as a characteristic. The principle behind the use of edge betweenness centrality is that a link with the highest edge betweenness centrality is likely to be a bottleneck link because a larger number of shortest paths travel this link than any other links. The algorithm categorizes virtual networks with the same bottleneck link, i.e., highest edge betweenness centrality link, into the same group, as illustrated in Fig. 4.6 and selects one representative virtual network from each group, thereby generating diverse virtual networks in terms of bottleneck links.

A formal definition of the virtual network classification is summarized as follows:

$$G_p = \{g_i \mid g_i \in G, C(g_i, l_p) = \max_q C(g_i, l_q)\}, \tag{4.6}$$

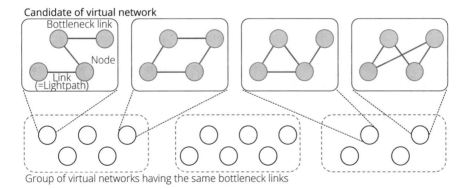

Candidate of virtual network
Group of virtual networks having the same bottleneck links

Fig. 4.6 Classification of virtual network candidates

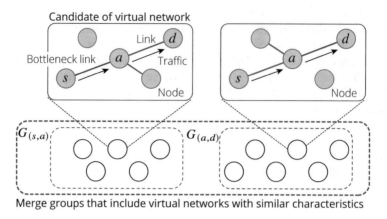

Merge groups that include virtual networks with similar characteristics

Fig. 4.7 Merging groups of similar virtual networks

where $p = (s, d)$ denotes a node pair of source node s and destination node d, l_p denotes a link (lightpath) established between node pair p, and $C(g_i, l_p)$ is the value of edge betweenness centrality of l_p in a virtual network g_i. Virtual networks in the group G_p are expected to contain the bottleneck link l_p.

The number of groups is still high compared to the requirement $0.1n^2$. Specifically, the number of groups is at most n^2 because the number of possible lightpaths is n^2. We must therefore reduce the number of groups further, and the algorithm consequently merges the virtual network groups according to the similarity of the location of bottleneck links. Figure 4.7 illustrates the condition of merging virtual network groups. Let us assume that a large amount of traffic flows from node s to d via a whose degree is low, and $l_{(s,a)}$ is the bottleneck of this virtual network. In this case, a large part of the traffic on link $l_{(s,a)}$ also flows on link $l_{(a,d)}$ because the degree of a is low. Thus, the load on $l_{(a,d)}$ is likely to be high. In this sense, virtual network candidates that belong to group $G_{(s,a)}$ and $G_{(a,d)}$ have similar characteristics. In the same way, virtual networks in $G_{(s,d)}$ also have similar characteristics. According to this observation, the algorithm merges the virtual network groups as

$$G_{(s,d)} \leftarrow G_{(s,a)} \cup G_{(a,d)} \cup G_{(s,d)}. \tag{4.7}$$

The algorithm selects three nodes, in this case, a, s, and d, in ascending order of their degree since the correlation of the traffic load on $l_{(s,a)}$ and $l_{(a,d)}$ is high in the case that the degree of a is low. Note that each group has many virtual networks, and the degree of a of the virtual networks differs depending on the network topologies. The algorithm thus selects the three nodes according to the degree of each node averaged over all the virtual networks in the group. The algorithm repeatedly merges virtual network groups, while the number of groups is greater than $0.1n^2$.

Step 3. Selection of Representative Virtual Networks Finally, the algorithm selects a virtual network whose maximum edge betweenness centrality is the lowest among other virtual networks in the group as the representative of each group. The rationale is that the maximum link utilization of a virtual network with low edge betweenness centrality is likely to also be low. The algorithm then embeds the selected virtual networks in the attractor structure.

4.4.4 Scalable Design of Attractor Structure by Graph Contraction

Although the proposed algorithm is able to design diverse attractors, as demonstrated in [22], it has scalability problems due to the high computational complexity of the generation process of isomorphic virtual networks. The number of isomorphic virtual networks for a given network is $n!$, and a commercial off-the-shelf computer, for instance, was able to compute the algorithm for physical networks with up to 10 nodes in a reasonable time in our experiment. This section thus extends the algorithm proposed in Sect. 4.4.3 to design attractors for large-scale physical networks.

The algorithm derives a graph minor of a given physical network to reduce the number of nodes in the physical network and performs the procedures described in Sect. 4.4.3 for the graph minor of the physical network. Instead of contracting edges, the algorithm divides the given physical network into clusters and places edges between the clusters, as illustrated in Fig. 4.8. The edge contraction procedure is outlined as follows:

Step 1   Divide a physical network into clusters.

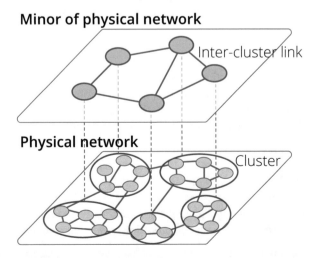

Fig. 4.8 Contraction of physical network topology

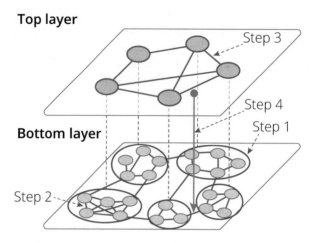

Fig. 4.9 Outline of hierarchical design of attractors

Step 2   Construct virtual network candidates in clusters at the bottom layer.
Step 3   Construct virtual network candidates at upper layers according to the
            procedures explained in Sect. 4.4.3.
Step 4   Connect lightpaths between clusters to nodes in the clusters.

Figure 4.9 illustrates this procedure. The remainder of this section explains the
procedure step by step.

Step 1. Cluster Division of a Physical Network The algorithm divides a given
physical network into c clusters. If the number of vertices in a cluster is greater
than c, the algorithm divides the cluster into further small clusters recursively until
the number of vertices in all clusters is equal to or less than c. This procedure implies
that the algorithm creates a hierarchical structure for a given physical network, as
illustrated in Fig. 4.9. An upper layer consists of clusters containing nodes in a lower
layer.

Step 2. Construction of Virtual Networks Inside Clusters at the Bottom Layer The
algorithm starts constructing virtual networks from the bottom layer. In the bottom
layer, the algorithm constructs virtual networks possessing either a full-mesh or a
star topology inside each cluster so that each topology in the clusters can adapt to
changes in a cluster and maintain connectivity in the case of network failure.

Step 3. Construction of Virtual Networks in Upper Layers The algorithm then
designs virtual networks in the upper layers according to the procedures described
in Sect. 4.4.3. It should be noted that the algorithm does not need to merge virtual
network groups. This step prepares at most $c(c - 1)/2$ virtual network groups
because the algorithm sets up lightpaths bidirectionally to reduce the number of
possible virtual networks. The detailed procedure is summarized as follows:

Step 3–1:   Construct a virtual network g using an existing heuristic method, i.e.,
              I-MLTDA [2], and generate isomorphic virtual networks from g.

Step 3–2:   Categorize the isomorphic virtual networks into at most $c(c-1)/2$
            groups on the basis of edge betweenness centrality.
Step 3–3:   Select the representative virtual network from each group.

Step 4. Establishment of Links Between Clusters The algorithm finally connects
clusters by creating lightpaths. This procedure corresponds to mapping lightpaths in
the upper layers to those in the bottom layer. The algorithm establishes lightpaths
between clusters from the kth layer to the $(k+1)$th layer, i.e., from an upper layer
to a lower layer. The probability of a lightpath $l_{i,j}^k$ being established between u and
v ($u \in V_i^k$ and $v \in V_j^k$) is formulated as follows:

$$P_{u,v} = (k_u k_v)^{-1}, \tag{4.8}$$

where $l_{i,j}^k$ represents a lightpath bidirectionally established between C_i^k and C_j^k, C_x^k
represents the xth cluster in the kth layer, k_u represents the number of lightpaths
connected to node u (the degree of u), and V_x^k represents nodes belonging to C_x^k.
Equation (4.8) aims at balancing the traffic load. Because a large amount of traffic
is likely to travel via a node with a high degree, the algorithm connects nodes whose
degrees are low.

The proposed algorithm described in Sects. 4.3 and 4.4 exhibits high adaptability
to both changes in traffic and network failures. Readers can refer to [16, 17, 22] for
detailed descriptions of the performance evaluation.

4.5 Related Work

Before concluding this chapter, we briefly introduce engineering-based approaches
for recovering from network failures in wavelength-routed WDM networks. Such
approaches can be classified into two categories: protection and restoration [32].

With protection approaches, a dedicated backup lightpath for each working
lightpath is reserved for recovery from network failures at network design time.
Protection approaches generally enable fast recovery from expected network failures [20, 26, 27]. Such approaches, however, cannot handle unexpected network
failures because they exploit several assumptions or prior knowledge about network
failures and predefine backup lightpaths at network design time. For instance, most
protection approaches do not take into account a situation in which both working
and backup lightpaths fail simultaneously. Though several protection approaches
that enable recovery from dual-link failures have been proposed [28], they also
exploit several assumptions or prior knowledge about network failures. Therefore,
protection approaches are generally unable to handle unexpected network failures.

In contrast, restoration approaches dynamically reconfigure an alternative virtual
network for lightpaths affected by a network failure when the failure occurs [4].
Restoration approaches discover spare resources in the network by collecting
network information, such as surviving optical fibers and OXCs, to establish an

alternative virtual network. Restoration approaches can maintain the connectivity of virtual networks if they are able to collect network information. However, if they are unable to collect network information changed by the failure, recovery from the failure is not possible.

Protection and restoration approaches, which are based on the control paradigm developed in the field of engineering, take into account a specific class of failures and are optimized to achieve fast and efficient recovery from the assumed failures. Hence, they may not be able to recover from unexpected network failures, such as multiple and series of network failures caused by a disaster. It is difficult for engineering approaches to adapt to various environmental changes as long as they use predefined algorithms. Therefore, the development of a virtual topology control method that adapts to various environmental changes in networks is indispensable.

4.6 Conclusion

This chapter presents a Yuragi-based virtual network control method, which is based on attractor selection found in a gene regulatory network, focusing on the similarity between gene regulatory networks and virtual network control. Unlike most engineering-based virtual network control methods, the proposed method does not rely on predefined algorithms. Instead, it exploits stochastic behavior in the case that the performance of a system is degraded, thereby adapting to unknown changes in the network environment. Since a system driven by attractor selection converges to one of the attractors embedded in its attractor structure, it is a challenge to design attractors to allow the system to adapt to various environmental changes. This chapter proposes two approaches for designing attractors.

Although attractor selection is applied to virtual network control in a wavelength-routed WDM network in this chapter, the proposed approach is applicable to other networks, such as virtualized networks and software-defined networks, which have recently attracted significant attention.

References

1. Arakawa, S., Murata, M., Miyahara, H.: Functional partitioning for multi-layer survivability in IP over WDM networks. IEICE Trans. Commun. **E83-B**(10), 2224–2233 (2000)
2. Banerjee, D., Mukherjee, B.: Wavelength-routed optical networks: Linear formulation, resource budgeting tradeoffs, and a reconfiguration study. IEEE/ACM Trans. Netw. **8**(5) (2000)
3. Baram, Y.: Orthogonal patterns in binary neural networks. Technical Memorandum 100060, NASA (1988)
4. Cheng, X., Shao, X., Wang, Y.: Multiple link failure recovery in survivable optical networks. Photon Netw. Commun. **14**(2), 159–164 (2007)
5. Chowdhury, N.M.M.K., Boutaba, R.: A survey of network virtualization. Comput. Netw. **54**, 862–876 (2010)

6. Comellas, J., Martinez, R., Prat, J., Sales, V., Junyent, G.: Integrated IP/WDM routing in GMPLS-based optical networks. IEEE Netw. Mag. **17**(2), 22–27 (2003)
7. Durán, R.J., Lorenzo, R.M., Merayo, N., de Miguel, I., Fernández, P., Aguado, J.C., Abril, E.J.: Efficient reconfiguration of logical topologies: Multiobjective design algorithm and adaptation policy. In: Proceedings of BROADNETS, pp. 544–551 (2008)
8. Furusawa, C., Kaneko, K.: A generic mechanism for adaptive growth rate regulation. PLoS Comput. Biol. **4**(1), e3 (2008)
9. Gençata, A., Mukherjee, B.: Virtual-topology adaptation for WDM mesh networks under dynamic traffic. IEEE/ACM Trans. Netw. **11**(2), 236–247 (2003)
10. Ghani, N., Dixit, S., Wang, T.S.: On IP-over-WDM integration. IEEE Commun. Mag. **38**(3), 72–84 (2000)
11. Hopfield, J.J.: Neural networks and physical systems with emergent collective computational abilities. Proc. Natl. Acad. Sci. U. S. A. **79**(8), 2554–2558 (1982)
12. Hopfield, J.J.: Neurons with graded response have collective computational properties like those of two-state neurons. Proc. Natl. Acad. Sci. U. S. A. **81**, 3088–3092 (1984)
13. Kaneko, K.: Life: an introduction to complex systems biology. In: Understanding Complex Systems. Springer, New York (2006)
14. Kashiwagi, A., Urabe, I., Kaneko, K., Yomo, T.: Adaptive response of a gene network to environmental changes by fitness-induced attractor selection. PLoS ONE **1**(1), e49 (2006)
15. Kodialam, M., Lakshman, T.V.: Integrated dynamic IP and wavelength routing in IP over WDM networks. In: Proceedings of IEEE INFOCOM, pp. 358–366 (2001)
16. Koizumi, Y., Miyamura, T., Arakawa, S., Oki, E., Shiomoto, K., Murata, M.: Adaptive virtual network topology control based on attractor selection. IEEE/OSA J. Lightwave Technol. **28**(11), 1720–1731 (2010)
17. Koizumi, Y., Arakawa, S., Murata, M.: Nature-inspired computing and optimization: theory and applications. In: Adaptive Virtual Topology Control Based on Attractor Selection. Springer, Berlin (2017)
18. Lee, K., Shayman, M.A.: Rollout algorithms for logical topology design and traffic grooming in multihop WDM networks. In: Proceedings of IEEE Global Telecommunications Conference 2005 (GLOBECOM '05), vol. 4 (2005)
19. Li, J., Mohan, G., Tien, E.C., Chua, K.C.: Dynamic routing with inaccurate link state information in integrated IP over WDM networks. Comput. Netw. **46**, 829–851 (2004)
20. Lin, T., Zhou, Z., Thulasiraman, K.: Logical topology survivability in IP-over-WDM networks: Survivable lightpath routing for maximum logical topology capacity and minimum spare capacity requirements. In: Proceedings of the International Workshop on the Design of Reliable Communication Networks (2011)
21. Mukherjee, B., Banerjee, D., Ramamurthy, S., Mukherjee, A.: Some principles for designing a wide-area WDM optical network. IEEE/ACM Trans. Netw. **4**(5), 684–696 (1996)
22. Ohba, T., Arakawa, S., Koizumi, Y., Murata, M.: Scalable design method of attractors in noise-induced virtual network topology control. IEEE/OSA J. Opt. Commun. Networking **7**, 851–863 (2015)
23. Rahman, Q., Sood, A., Aneja, Y., Bandyopadhyay, S., Jaekel, A.: Logical topology design for WDM networks using Tabu search. In: Distributed Computing and Networking. Lecture Notes in Computer Science, vol. 7129, pp. 424–427. Springer, Berlin (2012)
24. Ramaswami, R., Sivarajan, K.N.: Design of logical topologies for wavelength-routed optical networks. IEEE J. Sel. Areas Commun. **14**, 840–851 (1996)
25. Rojas, R.: Neural Networks: A Systematic Introduction. Springer, Berlin (1996)
26. Sahasrabuddhe, L., Ramamurthy, S., Mukherjee, B.: Fault management in IP-over-WDM networks: WDM protection versus IP restoration. IEEE J. Sel. Areas Commun. **20**(1), 21–33 (2002)
27. Shen, G., Grover, W.D.: Extending the *p*-cycle concept to path segment protection for span and node failure recovery. IEEE J. Sel. Areas Commun. **21**(8), 1306–1319 (2003)
28. Sivakumar, M., Sivalingam, K.M.: On surviving dual-link failures in path protected optical WDM mesh networks. Opt. Switch. Netw. **3**(2), 71–88 (2006)

29. Xin, Y., Rouskas, G.N., Perros, H.G.: On the physical and logical topology design of large-scale optical networks. IEEE/OSA J. Lightwave Technol. **21**(4), 904–915 (2003)
30. Ye, T., Zeng, Q., Su, Y., Leng, L., Wei, W., Zhang, Z., Guo, W., Jin, Y.: On-line integrated routing in dynamic multifiber IP/WDM networks. IEEE J. Sel. Areas Commun. **22**(9), 1681–1691 (2004)
31. Zhang, Y., Murata, M., Takagi, H., Ji, Y.: Traffic-based reconfiguration for logical topologies in large-scale WDM optical networks. J. Lightwave Technol. **23**, 1991–2000 (2005)
32. Zhou, D., Subramaniam, S.: Survivability in optical networks. IEEE Netw. **14**(6), 16–23 (2000)

Part II
Yuragi Learning: Extension to Artificial Intelligence

Chapter 5
Introduction to Yuragi Learning

Shin'ichi Arakawa and Tatsuya Otoshi

Abstract Yuragi learning is a paradigm that uses attractors as state templates of network systems. Under uncertain or fluctuating traffic in the network systems, template matching with predefined attractors is first examined, and the state of the network system is then changed to the most appropriate attractors. In this introductory chapter to Yuragi learning and its applications, we present a fundamental model of human perceptual decision-making, which is the core of Yuragi learning.

5.1 Yuragi Learning: An Introduction

As discussed in Chap. 1, Yuragi theory and Yuragi control enable the adaptive control of network systems. Attractors characterize the deterministic behavior of the control and are selected through the activity and Yuragi. Although the trial-and-error paradigm driven by Yuragi achieves adaptiveness, many trials may be required to find a favorable solution. From an engineering perspective, this is not preferable, as a large number of trials require a long time to adapt and disrupt application services over the network system.

One way to eliminate the trial-and-error paradigm in network control is to change the role of the attractors. In Yuragi control, attractors are a type of landmark in dynamical systems that require a trial-and-error paradigm to reach a system state at one of the landmarks. An alternative paradigm is to use attractors as state templates of the network systems. Specifically, under uncertain or fluctuating traffic in network systems, template matching with predefined attractors is first examined, and the state of the network systems is then changed to the most appropriate attractor. This type of paradigm is also observed in the human brain and is known as *decision-making*.

S. Arakawa (✉)
Graduate School of Information Science and Technology, Osaka University, Suita, Osaka, Japan
e-mail: arakawa@ist.osaka-u.ac.jp

T. Otoshi
Graduate School of Economics, Osaka University, Toyonaka, Osaka, Japan
e-mail: t-otoshi@econ.osaka-u.ac.jp

© Springer Nature Singapore Pte Ltd. 2021
M. Murata, K. Leibnitz (eds.), *Fluctuation-Induced Network Control and Learning*,
https://doi.org/10.1007/978-981-33-4976-6_5

Decision-making in the human brain is the most familiar example of decision-making under uncertainty. In daily life, the human brain estimates various scenarios using sensory stimuli from the environment and decides which actions to take. Thus, introducing this brain mechanism into network control is a promising approach for handling uncertain observations to determine the appropriate network state. The current consensus in cognitive science states that the brain accumulates sensory information over a period of time, and makes a perceptual decision (i.e., categorizes observed information) when sufficient information has been collected [2, 4, 6].

In this chapter, we first present the fundamentals of the Bayesian attractor model (BAM) [2] as a model of human perceptual decision-making. Then, we introduce *Yuragi learning*, where the state of network systems is controlled with the assistance of the BAM.

5.2 Bayesian Attractor Model for Human Perceptual Decision-Making

5.2.1 Overview

The BAM models the behavior of the human brain. For example, when a traffic light turns green at an intersection, we see a green light and make the decision to proceed. More precisely, the color categories of traffic lights are retained in our brains, and we judge which category the observed sensory information belongs to. Even when the sensory information contains a large amount of noise (e.g., when it is difficult to see a traffic light due to bad weather or backlight), the brain is able to accumulate evidence extracted from this noisy sensory information over time, and makes the appropriate perceptual decisions.

The BAM has a state variable $\mathbf{z}$ that eventually settles to a fixed point ϕ that is defined by the attractor dynamics (i.e., winner-take-all dynamics [7]) as evidence is accumulated. Internally, the BAM has several fixed points ϕ_i, each of which corresponds to a choice for the long-term average $\boldsymbol{\mu}_i$ of an observed value. At time t, the model infers the posterior distribution of the state variable $\mathbf{z}_t$, denoted by $p(\mathbf{z}_t|\mathbf{X}_{1:t})$, given observations up to time t, denoted by $\mathbf{X}_{1:t} = \{\mathbf{x}_1, \cdots, \mathbf{x}_t\}$. The model selects $\boldsymbol{\mu}_i$ as soon as a confidence criterion, such as

$$ p(\mathbf{z}_t = \phi_i|\mathbf{X}_{1:t}) \geq \lambda, $$

is satisfied. Here, the posterior belief $p(\mathbf{z}_t = \phi_i|\mathbf{X}_{1:t})$ is the confidence measure for making a decision for choice $\boldsymbol{\mu}_i$. Thus, the BAM accumulates observation values and makes a decision when the confidence for the decision is sufficiently large.

5.2.2 Inference Mechanism for Decision-Making by Bayesian Attractor Model

The BAM has a generative model for Bayesian inference by the decision maker (i.e., brain). The generative model calculates the likelihood of observations for all possible scenarios considered by the decision maker. More precisely, the generative model predicts a probability distribution over observation values based on the current state variable and its attractor dynamics.

The generative model defines a change in the state variable from one-time step to the next as

$$\mathbf{z}_t - \mathbf{z}_{t-\Delta t} = \Delta t \cdot f(\mathbf{z}_{t-\Delta t}) + \sqrt{\Delta t} \cdot \mathbf{w}_t, \qquad (5.1)$$

where $\mathbf{z}_t$ is the D-dimensional state variable at time t, and $f(\mathbf{z})$ is the attractor dynamics [7].

The noise term $\mathbf{w}_t$ follows the normal distribution $N(\mathbf{0}, \mathbf{Q})$, where $\mathbf{Q} = (q^2/\Delta t) \cdot \mathbf{I}$ is the variance–covariance matrix of the noise, and q represents the dynamic uncertainty. The dynamic uncertainty represents the amount of noise with which the decision maker expects the state variable to be changed, which is interpreted as the tendency for state variables to switch between fixed points.

The generative model predicts a probability distribution over the observation values, given the state variable $\mathbf{z}$. The equation for the prediction is

$$\mathbf{x} = \mathbf{M} \cdot \sigma(\mathbf{z}) + \mathbf{v} \qquad (5.2)$$
$$= [\boldsymbol{\mu}_1, \cdots, \boldsymbol{\mu}_D] \cdot \sigma(\mathbf{z}) + \mathbf{v}$$
$$= \sigma(z_1) \cdot \boldsymbol{\mu}_1 + \sigma(z_2) \cdot \boldsymbol{\mu}_2 + \cdots + \sigma(z_D) \cdot \boldsymbol{\mu}_D + \mathbf{v},$$

where $\mathbf{M} = [\boldsymbol{\mu}_1, \cdots, \boldsymbol{\mu}_D]$ contains the averages of observation values that correspond to choices and $\sigma(\mathbf{z})$ is a sigmoid function that maps all $z_j \in \mathbf{z}$ to values between 0 and 1. Due to the winner-take-all dynamics of $\mathbf{z}$, the fixed point ϕ_i is mapped to a vector $\sigma(\phi_i)$, where one element is approximately 1 and the other elements are approximately 0. Thus, the linear combination $\mathbf{M} \cdot \sigma(\mathbf{z})$ associates each fixed point ϕ_i with the choice (average of observations) $\boldsymbol{\mu}_i$. The noise term $\mathbf{v}$ follows the normal distribution $N(\mathbf{0}, \mathbf{R})$, where $\mathbf{R} = r^2 \cdot \mathbf{I}$ is the variance–covariance matrix of the noise, and r represents the sensory uncertainty. The sensory uncertainty represents the amount of noise in observations that the decision maker expects. In contrast, the actual amount of noise by which observations deviate from the average values is denoted by s. We summarize the key parameters of the BAM in Table 5.1.

At time t, the BAM infers the posterior distribution of the state variable $\mathbf{z}_t$, denoted by $p(\mathbf{z}_t|\mathbf{X}_{1:t})$, using the generative model and the unscented Kalman filter (UKF) [5]. The UKF is a statistical sampling method that approximates the posterior distribution $p(\mathbf{z}_t|\mathbf{X}_{1:t})$ with a normal distribution. In the following, we

Table 5.1 Key parameters of Bayesian attractor model

Parameter	Explanation
s (noise level)	Actual amount of noise in observation values
q (dynamic uncertainty)	Tendency for state variables to switch between fixed points
r (sensory uncertainty)	Amount of noise in observation values that the decision maker expects

briefly describe the flow of the Bayesian inference in the BAM. First, the generative model predicts the posterior distribution of the state variable at time t using Eq. (5.1) and approximates it with a normal distribution $N(\hat{\mathbf{z}}_t, \hat{\mathbf{P}}_t)$, where $\hat{\mathbf{P}}_t$ represents the variance–covariance matrix of the predicted state variable, $\hat{\mathbf{z}}_t$. Second, the generative model predicts the probability distribution of the corresponding observation values using Eq. (5.2) and approximates it with a normal distribution $N(\hat{\mathbf{x}}_t, \hat{\mathbf{\Sigma}}_t)$, where $\hat{\mathbf{\Sigma}}_t$ represents the variance–covariance matrix of the predicted observation values, $\hat{\mathbf{x}}_t$. Finally, the BAM calculates the observation residual between the predicted observation values $\hat{\mathbf{x}}_t$ and the actual observation values $\mathbf{x}_t$,

$$\epsilon_t = \mathbf{x}_t - \hat{\mathbf{x}}_t, \tag{5.3}$$

and updates the estimate of the state variable $\bar{\mathbf{z}}_t$ and its posterior variance–covariance matrix $\bar{\mathbf{P}}_t$ via the Kalman gain $\mathbf{K}_t$ as follows:

$$\bar{\mathbf{z}}_t = \hat{\mathbf{z}}_t + \mathbf{K}_t \cdot \epsilon_t,$$

$$\bar{\mathbf{P}}_t = \hat{\mathbf{P}}_t - \mathbf{K}_t \hat{\mathbf{C}}_t^T.$$

The Kalman gain represents the relative importance of the observation residual and is given by

$$\mathbf{K}_t = \hat{\mathbf{C}}_t \hat{\mathbf{\Sigma}}_t^{-1},$$

where $\hat{\mathbf{C}}_t$ is the covariance matrix between the predicted state variable $\hat{\mathbf{z}}_t$ and the predicted observation values $\hat{\mathbf{x}}_t$. In this way, the BAM approximates the posterior distribution of the state variable $p(\mathbf{z}_t | \mathbf{X}_{1:t})$ with a normal distribution $N(\bar{\mathbf{z}}_t, \bar{\mathbf{P}}_t)$.

5.2.3   Design Choices for Bayesian-Attractor-Model-Based Network Control

In [2], the authors examined the behavior of the BAM, focusing on two-alternative forced choice tasks, which are commonly employed when investigating perceptual decision-making. Specifically, the authors considered random dot motion tasks that identify the direction in which a randomly moving cloud of dots moves on average. Several problems must be solved in order to apply the BAM-based approach to network control, which usually involves many alternatives.

5.2.3.1 Setting Parameters r and q

Parameter r denotes the sensory uncertainty, which represents the amount of noise in observations that the decision maker expects. Thus, r should be the empirical standard deviation s of observations. In fact, it was demonstrated in [2] that the optimal Bayesian decision maker should have a generative model in which r is ideally equal to s.

Parameter q denotes the dynamic uncertainty, which controls the propensity of decision makers to change their decision and affects the balance between flexibility and stability in decision-making. When q is small, the state variable $\mathbf{z}$ is too stable to switch between fixed points. That is, it is difficult for decision makers to change their decision even when the actual choice (i.e., average of observation values) changes, since they dismiss evidence for another choice as noise. When q is large, although decision makers can change their decision rapidly, they sometimes change their decision due to sensory noise. Therefore, it is necessary to set q to an appropriate value with which decision makers can make fast and accurate decisions. By examining the effects of parameters r and q on the BAM-based approach in advance by offline simulations, we can obtain appropriate parameter sets $\{r, q\}$.

5.2.3.2 How to Determine a Criterion for Decision-Making

The criterion for decision-making should be suitable for a VN (virtual network) reconfiguration method. In [1], the authors introduced several definitions of confidence for decision-making: the posterior belief itself, the logarithm of the posterior belief, and the log change in the posterior belief.

5.2.3.3 Preparing Attractors

In the BAM, it is crucial to design attractors properly, as attractors represent state templates in network control. An extreme approach is to prepare all network states as attractors. However, this requires a long recognition time. Thus, a limited number of attractors should be prepared in the BAM.

One problem is to determine how to design attractors in BAM-based network control and how and when to update attractors in dynamic systems. However, there are no clear guidelines for designing or updating attractors, and this problem must be investigated separately in each application. A possible guideline can be obtained from a form of Eq. (5.2). The first term of the right-hand side of the equation forms a linear combination; thus, the attractors should have linear independence, and the dimension of the space spanned by attractors becomes maximal.

5.3 Virtual Network Reconfiguration Based on Yuragi Learning

5.3.1 Overview of Virtual Network Reconfiguration

Figure 5.1 illustrates how to apply the BAM-based approach to the VN reconfiguration problem. When the traffic situation is identified as μ_i given the observation values $\mathbf{X}_{1:t}$, the corresponding VN g_i suitable for the identified traffic situation μ_i is retrieved and configured. More precisely, we select VN g_i when the confidence in decision-making for the choice μ_i is sufficiently large. Here, we use the amount of incoming and outgoing traffic at edge routers as the observation values $\mathbf{X}_{1:t}$. Note that the VN g_i is prepared in advance such that the VN is suitable for the traffic situation μ_i.

Applying only the BAM-based approach is insufficient for a VN reconfiguration framework because the retrieved VN g_i may not be able to accommodate the traffic demand even when the identification succeeds. To handle this case, we incorporate the Yuragi control discussed in Chap. 4 into our VN reconfiguration framework to find favorable VNs. That is, we prepare a set of control phases and change the control phase based on both the confidence from the BAM-based approach and the service quality of the VN. More precisely, our VN reconfiguration framework is an online algorithm that reconfigures a VN by the following steps based on the observation of the amount of incoming and outgoing traffic at edge routers and the service quality of the VN (here, the maximum link utilization in the VN).

(Step 1) Calculate the confidence of the BAM-based approach using the measured amount of incoming and outgoing traffic at edge routers.

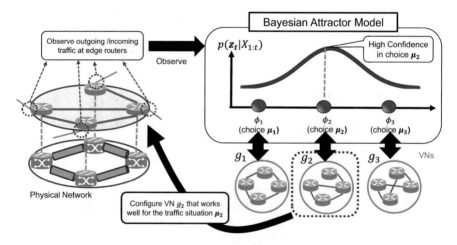

Fig. 5.1 Application of Bayesian attractor model to virtual network (VN) reconfiguration

(Step 2)   Change the control phase based on the confidence of the BAM-based approach and the service quality of the VN, and execute the control.

Note that the amount of incoming and outgoing traffic at edge routers and the link utilization in the VN can be retrieved more easily than information of the end-to-end traffic demand matrix. The control phases are as follows:
(Phase 1) Stay until a new traffic situation is identified.

- We do not reconfigure a VN until the current traffic situation is identified.

(Phase 2) Reconfigure the VN based on the identified traffic situation.

- We select the VN candidate g_i that is suitable for the identified traffic situation μ_i (Phase 2-1).
- If VN g_i cannot accommodate the traffic demand, we search for a suitable VN using Yuragi control (Phase 2-2).

In summary, our VN reconfiguration framework first identifies traffic situations using the BAM-based approach (Phase 1) and immediately changes the VN after successful identification (Phase 2-1). Then, we measure the service quality of the VN and reconfigure the VN if necessary (Phase 2-2). Note that this VN reconfiguration framework does not cover the case in which the identification of traffic situations fails, that is, where the confidence is stable at a small value. We investigate ways to handle the case in which the identification fails in Sect. 5.5.

5.3.2   Virtual Network Reconfiguration Algorithm

The details of the VN reconfiguration framework are presented in the following subsections.

5.3.2.1   Preparation

We prepare VNs g_i, $(1 \leq i \leq D)$, that are well suited for traffic situations μ_i in advance. Examples of VN preparation include the following:

- Extracting VNs from the control history of VN configurations that indicate adequate performance in specific traffic situations.
- Calculating VNs using traffic demand matrices that can be predicted from past traffic fluctuations.

We also prepare sets of the parameters $\{r, q\}$ of the BAM with which the traffic situation can be successfully identified by offline simulations.

5.3.2.2 (Step 1) Calculate Confidence Using the BAM-Based Approach

At time t, we observe the amount of incoming and outgoing traffic at edge routers and calculate the confidence in decision-making for the various choices.

First, we determine the parameters r and q used for inference. Specifically, we calculate the empirical standard deviation s_t using the observations up to time t, $\mathbf{X}_{1:t}$, and set r to s_t. Then, we set q to the value corresponding to r obtained in the above preparatory phase. Here, we sequentially update the empirical standard deviation s_t using Welford's method [8], as it is not necessary to retain past observation values from time 1 to time $t - 1$. Second, we infer the posterior distribution of the state variable, $p(\mathbf{z}_t|\mathbf{X}_{1:t})$, and calculate the posterior belief for each choice, $p(\mathbf{z}_t = \phi_i|\mathbf{X}_{1:t})$. Finally, we calculate the confidence of the decision that the current traffic situation is identified as $\boldsymbol{\mu}_i$. Here, following the method in [1], we use as the confidence the left-hand side of Eq. (5.4), which represents the difference between the logarithms of the posterior beliefs. That is, we identify the current traffic situation as $\boldsymbol{\mu}_i$ when

$$\log_{10} \frac{p(\mathbf{z}_t = \phi_i|\mathbf{X}_{1:t})}{p(\mathbf{z}_t = \phi_j|\mathbf{X}_{1:t})} \geq \lambda, \tag{5.4}$$

where the posterior belief in choice $\boldsymbol{\mu}_i$, $p(\mathbf{z}_t = \phi_i|\mathbf{X}_{1:t})$, is the largest among all the choices, and the posterior belief in choice $\boldsymbol{\mu}_j$, $p(\mathbf{z}_t = \phi_j|\mathbf{X}_{1:t})$, is the second largest.

5.3.2.3 (Step 2) Change the Control Phase and Execute the Control

We change the control phase based on the confidence obtained in Step 1 and the service quality of the VN and execute the control. The state transition diagram of the control phases is presented in Fig. 5.2. The label at each edge represents the transition condition, which consists of the confidence from the BAM-based approach and the service quality of the VN. The confidence becomes stable at a high value when Eq. (5.4) is satisfied for a total of c consecutive times. The details of each control phase are as follows:

(Phase 1) Stay until a new traffic situation is identified.

- This phase is the initial state of our VN reconfiguration framework. The control phase also changes to this phase when the confidence falls below threshold λ.
- We do not reconfigure a VN in this phase.

(Phase 2-1) Configure the VN candidate that is well suited for the identified traffic situation.

- The control phase changes from Phase 1 to this phase when the confidence becomes stable at a high value.

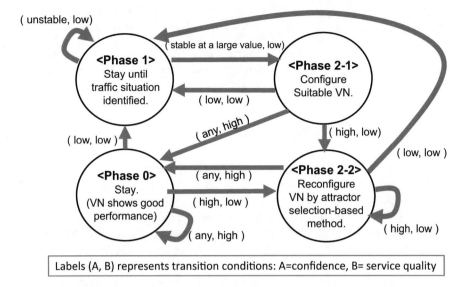

Fig. 5.2 State transition diagram for the virtual network reconfiguration framework

- We configure one of VN candidates g_i that is well suited for the traffic situation μ_i in this phase.

(Phase 2-2) Reconfigure the VN using the attractor-selection-based method.

- The control phase changes to this phase when the confidence takes a high value and the service quality of the VN is low.
- We search for suitable VNs using Yuragi control in this phase.

(Phase 0) Stay (because the VN shows favorable performance).

- The control phase changes to this phase when the service quality on the VN improves.
- We do not reconfigure the VN in this phase.

5.4 Performance Evaluation of Yuragi Learning

In this section, we first investigate the characteristics of our VN reconfiguration framework based on Yuragi learning and then evaluate the advantages of our VN reconfiguration framework when the traffic demand fluctuates. We use VN reconfiguration over an elastic optical network for evaluation. Here, we evaluate the advantages of our VN reconfiguration framework by comparing it with an approach using Yuragi control (called the reference method hereafter), as neither method uses traffic demand matrix information.

Table 5.2 Parameters of the
USNET physical network
topology

Parameter	Value
Number of nodes	24
Number of links	43
Number of bandwidth-variable transceivers	10
Bandwidth of each BVT	100 Gbps

5.4.1   Evaluation Environments

We used an elastic optical network that had the USNET topology, and the parameters
of the physical network topology are provided in Table 5.2. The number of nodes,
each of which consisted of an IP router and a bandwidth-variable wavelength cross-
connect (BV WXC), was 24, and the number of links (i.e., bidirectional optical
fibers) was 43. Here, all IP routers were edge routers. Each BV WXC had 10
bandwidth-variable transponders (BVTs) that offered up to 100 Gbps of bandwidth.
The goal of the control was to cause the maximum link utilization in a VN to be
less than 0.5. We generated traffic demand matrices $T_1, \cdots, T_5$ that followed a
log-normal distribution. We denoted the amount of incoming and outgoing traffic at
edge routers $E_1, \cdots, E_5$, which were predefined traffic situations $\mu_1, \cdots, \mu_5$, when
the traffic demand matrices were $T_1, \cdots, T_5$, and calculated the configurations of
VN candidates $g_1, \cdots, g_5$, which could accommodate $T_1, \cdots, T_5$. It should be
noted that the predefined traffic situations are attractors in Yuragi learning. The VN
candidates were determined by the virtual topology using the most subcarriers first
algorithm[3], and frequency slots were allocated to lightpaths using the first–last fit
algorithm[11]. We set c to 3 and λ to 10.

For the evaluation, at every unit time step, the end-to-end traffic demand matrix
was generated based on the normal distribution $N(T_1, \Sigma)$ until time 100; then, it
was generated based on the normal distribution $N(T_2, \Sigma)$ until time 200, where
$T_i = (T_{i,11}, \cdots, T_{i,NN})$ and $\Sigma = CV^2\text{diag}(T_{i,11}^2, \cdots, T_{i,NN}^2)$. N denotes the
number of nodes, and CV is the coefficient of variation that represents the degree
of traffic fluctuation. In [9], the authors analyzed real traffic data and fit the data to
a traffic fluctuation model that followed a normal distribution. The results indicated
that the CV of real traffic is approximately within the range [0.5, 1.5]. The reference
method changed its control phase based on the service quality of the VN, as
illustrated in Fig. 5.3. The method searched for desirable VNs in Phase 2-2 and
then changed to Phase 0 when the performance of the VN improved.

5.4.2   Characteristics of Virtual Network Reconfiguration Framework

Figure 5.4 presents the transitions between control phases for each method when
CV was 0.5 and demonstrates the behavior of our framework. Our method started in
Phase 1 and remained in Phase 1 until the traffic situation was identified. At time 5,

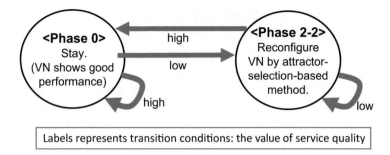

Labels represents transition conditions: the value of service quality

Fig. 5.3 State transition diagram for the attractor-selection-based method

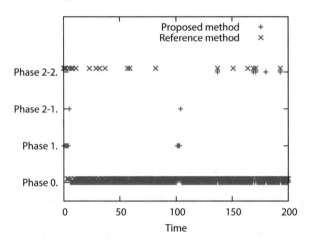

Fig. 5.4 Transition of the control phase ($CV = 0.5$)

the method shifted to Phase 2-1 and reconfigured the VN to the most promising VN among the predefined VN candidates. Thereafter, the method entered Phase 0, as the VN was able to accommodate the traffic. Because the confidence in the identification of the current traffic situation decreased at time 101, the method returned to Phase 1. At time 104, the method shifted to Phase 2-1, having detected a change in the traffic situation. It should be noted that our method was able to detect a change in the traffic situation even though it accumulated past observations. Thereafter, the method returned to Phase 0 because the VN was able to accommodate the traffic. Even when the method shifted to Phase 2-2 due to a traffic fluctuation at time 137, it immediately returned to Phase 0 by the attractor-selection-based method. This is because the VN was reconfigured to the most promising candidate, and therefore, the attractor-selection-based method required little effort to identify an effective VN. In contrast, the reference method repeatedly changed the control phase between Phase 2-2 and Phase 0. That is, although the reference method searched for a suitable VN for the current traffic situation and temporarily identified one, the VN was unable to adapt to traffic changes after that.

5.4.3 Advantages of Virtual Network Reconfiguration Framework

In this section, we evaluate the advantages of our VN reconfiguration framework. Our VN reconfiguration framework aims to detect changes in the traffic situation and configure a VN that is suitable for the current traffic situation. We therefore evaluated the advantages of our framework by considering the performance after the traffic situation changed (i.e., after time 100). We executed 100 trials with different seeds to generate different traffic demand matrices.

First, we evaluated the time required for identification of the traffic situation, and the distribution is presented in Fig. 5.5. Here, the required time is defined as the time from when the traffic situation changed (i.e., time 100) to the time when the confidence became stable at a high value and the control phase changed to Phase 2-1 for the first time. Figure 5.5 demonstrates that our VN reconfiguration framework identified the traffic situation within 10 time steps in all trials. It should also be noted that our framework identified the current traffic situation as μ_2 in all trials. That is, our framework correctly identified the traffic situation because the traffic demand matrix was generated by following $N(\mathbf{T}_2, \Sigma)$. We also observed that the required time increased as CV increased. This is because it took a longer time to accumulate evidence for a choice when the traffic demand fluctuated more and the observations were more likely to deviate from the attractors.

Second, we evaluated the performance of the VN configured in Phase 2-1. Figure 5.6 illustrates the distribution of the maximum link utilization of the VN configured in Phase 2-1 after the traffic situation changed (i.e., after time 100). It can be seen that in most trials, the VN configured in Phase 2-1 achieved the control goal of keeping the maximum link utilization less than 0.5. However, the maximum link utilization of the VN tended to increase as CV increased. That is, although our VN reconfiguration framework was able to adapt to changes in traffic situations by configuring the most suitable VN among the candidates in Phase 2-1 when the

Fig. 5.5 Cumulative distribution of the time required to identify the traffic situation

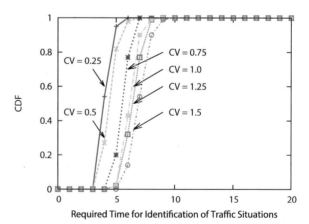

Fig. 5.6 Distribution of the
maximum link utilization of
the virtual network
configured in Phase 2-1

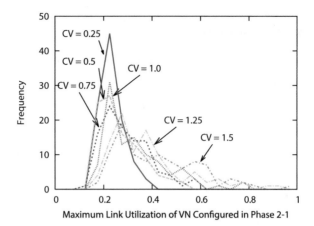

traffic fluctuation was small, the VN reconfiguration framework had to search for a
suitable VN in Phase 2-2 when the traffic fluctuation was large.

5.4.4 Impact of the Number of Attractors

In this section, we discuss the impact of the number of attractors (i.e., number of
predefined traffic situations) on the network control performance. We expected that
our proposed VN reconfiguration framework would be able to adapt to a larger
number of traffic situations when it could identify them. However, it was likely to
be more difficult to distinguish traffic situations as their number increased. Thus, we
evaluated the effect of the number of choices on our VN reconfiguration framework.

We denoted the number of attractors D, which is the evaluation parameter in this
section. That is, we generated traffic demand matrices $\mathbf{T}_1, \cdots, \mathbf{T}_D$ that followed
a log-normal distribution. Then, we denoted the amount of incoming and outgoing
traffic at edge routers by $\mathbf{E}_1, \cdots, \mathbf{E}_D$, which was the traffic situation $\mu_1, \cdots, \mu_D$,
when the traffic demand matrix was $\mathbf{T}_1, \cdots, \mathbf{T}_D$, and calculated configurations
of the VN candidates $g_1, \cdots, g_D$ that could accommodate the traffic demand
$\mathbf{T}_1, \cdots, \mathbf{T}_D$. We calculated the configuration of these VN candidates in the same
way as in Sect. 5.4.1. For the evaluation, CV was 0.5, and the other parameters were
identical to those in Sect. 5.4.1.

We considered the maximum time taken to identify the traffic situation for
various numbers of choices. Figure 5.7 presents the maximum time required to
identify the traffic situation over 100 trials with different seeds used to generate
different traffic demand matrices. It can be seen that it took longer to identify
traffic situations as their number increased. Specifically, although the required time
was at most approximately 10 time steps when the number of choices was 15 or
less, the required time increased when the number of choices became larger than
20. Therefore, there is a trade-off between the ability to identify a larger number

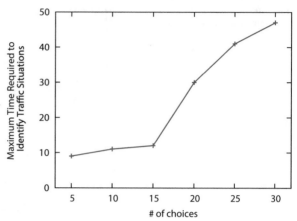

Fig. 5.7 Maximum time required to identify the traffic situation

of traffic situations and the ability to adapt to traffic changes in a shorter time period. It is thus desirable to limit the number of attractors to effectively deploy our VN reconfiguration framework and accommodate changing traffic demand. In this evaluation environment, the limit is 15.

5.5 Yuragi Learning with Linear Regression

The VN reconfiguration framework discussed in Sect. 5.3 can be extended to handle the case in which the identification of traffic situations fails. When the identification of traffic situations by the BAM-based method fails, we can calculate and configure a new VN using linear regression [10] by following the steps below. Figure 5.8 presents an outline of the VN reconfiguration method based on the BAM with linear regression.

(Step 1)   Fit the current traffic situation, denoted by μ_{new}, by linear regression using attractors $\mu_1, \cdots, \mu_D$.

(Step 2)   Calculate and configure a new VN, denoted by g_{new}, using the obtained regression coefficients $w_1, \cdots, w_D$.

Applying only the linear-regression-based approach is not sufficient when the identification of traffic situations fails because the VN g_{new} may not be able to accommodate the traffic demand. In this case, we apply Yuragi control to identify a suitable VN. However, by introducing the linear-regression-based method, it is expected that the number of VN reconfigurations to accommodate the traffic demand can be reduced even when the identification of traffic situations fails, as we calculate a suitable VN g_{new} using the information obtained by fitting the current traffic situation. Note that because this linear-regression-based method utilizes only the information used in the above VN reconfiguration framework, such as the observed amount of incoming and outgoing traffic at edge routers and attractors

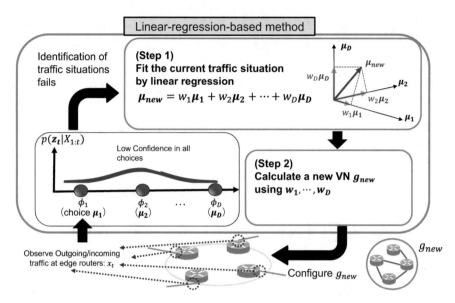

Fig. 5.8 Outline of the virtual network reconfiguration based on the Bayesian attractor model with linear regression

$(\boldsymbol{\mu}_1, \cdots, \boldsymbol{\mu}_D)$, there is no information to be additionally observed to configure the VN g_{new}. We incorporate this linear-regression-based method into the VN reconfiguration framework proposed in Sect. 5.3.

5.5.1 Virtual Network Reconfiguration Algorithm with Linear Regression

The details of the linear-regression-based VN reconfiguration algorithm are presented in the following subsections.

5.5.1.1 (Step 1) Fit the Traffic Situation by Linear Regression

The BAM as a state-space representation has the output equation Eq. (5.2), where the observation value (observed amount of incoming and outgoing traffic at edge routers in our VN reconfiguration framework) is represented by a linear sum of the predefined traffic situations of attractors. We apply this equation to handle the case in which the identification of traffic situations fails. That is, we fit the current traffic situation $\boldsymbol{\mu}_{new}$ by linear regression using attractors $\boldsymbol{\mu}_1, \cdots, \boldsymbol{\mu}_D$ to satisfy the equation

$$\boldsymbol{\mu}_{new} = w_1 \boldsymbol{\mu}_1 + \cdots + w_D \boldsymbol{\mu}_D + \boldsymbol{\epsilon} = \sum_{i=1}^{D} w_i \boldsymbol{\mu}_i + \boldsymbol{\epsilon}, \tag{5.5}$$

where $w_1, \cdots, w_D$ are the regression coefficients, and $\boldsymbol{\epsilon}$ represents the error term. More precisely, we calculate regression coefficients $w_1, \cdots, w_D$ so that the residual sum of squares (RSS) defined by the equation,

$$RSS(w_i) = \boldsymbol{\epsilon}^T \boldsymbol{\epsilon} = (\boldsymbol{\mu}_{new} - \hat{\boldsymbol{\mu}}_{new})^T (\boldsymbol{\mu}_{new} - \hat{\boldsymbol{\mu}}_{new}) \tag{5.6}$$

$$= \left(\boldsymbol{\mu}_{new} - \sum_{i=1}^{D} w_i \boldsymbol{\mu}_i \right)^T \left(\boldsymbol{\mu}_{new} - \sum_{i=1}^{D} w_i \boldsymbol{\mu}_i \right),$$

is minimized by the least squares method. In the BAM-based approach, the coefficient of each traffic situation $\boldsymbol{\mu}_i$, which is denoted by $w_i' = \sigma(z_i)$, is defined to satisfy $0 \leq w_i' \leq 1$. However, there are no constraints on coefficient w_i when we fit the current traffic situation $\boldsymbol{\mu}_{new}$ by linear regression. Even when the traffic volume increases and coefficient w_i is thereby greater than 1, it does not become more difficult to calculate a new VN g_{new}, as we can assume that the traffic pattern, which is the relationship between the traffic volume at every edge router, does not change. It should be noted that in this study, we do not consider the case in which the traffic volume increases such that the physical network equipment must be enhanced to accommodate the traffic demand.

5.5.1.2 (Step 2) Calculate a New VN

Utilizing the obtained regression coefficients $w_1, \cdots, w_D$, we calculate a new VN g_{new} and configure it. In our algorithm, because the current traffic situation $\boldsymbol{\mu}_{new}$ is fitted by a linear sum of attractors $\boldsymbol{\mu}_1, \cdots, \boldsymbol{\mu}_D$, we represent the current traffic demand matrix $\mathbf{T}_{new}$ as a linear sum of the corresponding traffic demand matrices $\mathbf{T}_1, \cdots, \mathbf{T}_D$, as expressed in following equation:

$$\mathbf{T}_{new} = w_1 \mathbf{T}_1 + \cdots + w_D \mathbf{T}_D = \sum_{i=1}^{D} w_i \mathbf{T}_i. \tag{5.7}$$

We can use the information of the traffic demand matrices $\mathbf{T}_1, \cdots, \mathbf{T}_D$ because we retain this information to calculate VN candidates in the BAM-based method. Then, we calculate a new VN g_{new} by heuristic algorithms [3, 11] using the obtained traffic demand matrix $\mathbf{T}_{new}$ as an input. When g_{new} cannot accommodate the traffic demand, we apply Yuragi control to find a suitable VN.

5.5.2 Effect of Linear Regression

We evaluated the effectiveness of configuring a new VN g_{new} calculated by the linear-regression-based method when the identification of traffic situation failed.

The parameters of the physical network topology, the control goal, and the information retained by the BAM-based method (e.g., the predefined traffic demand matrices $\mathbf{T}_1, \cdots, \mathbf{T}_5$, the corresponding traffic situations $\boldsymbol{\mu}_1, \cdots, \boldsymbol{\mu}_5$, and VN candidates $g_1, \cdots, g_5$) were the same as described in Sect. 5.4.1. We generated 1000 patterns of traffic demand matrix information $\mathbf{T}'$ with different seeds assuming unknown traffic fluctuations. It should be noted that the generated traffic demand matrices $\mathbf{T}'$ were different from the retained ones, $\mathbf{T}_1, \cdots, \mathbf{T}_5$. For the evaluation, at every unit time step, the end-to-end traffic demand matrix was generated based on the normal distribution $N(\mathbf{T}_1, \Sigma)$ until time 50; then, traffic fluctuation occurred, and the traffic demand matrix was generated based on the normal distribution $N(\mathbf{T}', \Sigma)$ until time 100, where $\mathbf{T}_i = (T_{i,11}, \cdots, T_{i,NN})$ and $\Sigma = 0.5^2 \mathrm{diag}(T_{i,11}^2, \cdots, T_{i,NN}^2)$.

Figure 5.9 presents the details of the simulation results, indicating whether the identification of traffic situations succeeded, and whether the configured VN was able to accommodate traffic demand for 1000 trials using different $\mathbf{T}'$. Figure 5.9 indicates that the identification of traffic situations succeeded in $30.9(= 12.5 + 18.4)\%$ of trials and failed in 69.1% of trials. To investigate the effectiveness of the linear-regression-based method when the identification of traffic situations failed, we focus on the latter 69.1% of the trials below. Figure 5.9 demonstrates that the configured VN was able to accommodate traffic demand in 65.7% of trials when the identification of traffic situations failed, whereas it was unable to accommodate traffic demand in 3.4% of trials. Thus, the linear-regression-based method was able

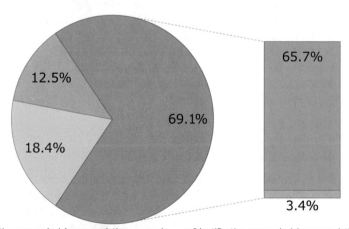

□ Identification succeeds / Accommodation succeeds   ■ Identification succeeds / Accommodation fails
■ Identification fails / Accommodation succeeds   ■ Identification fails / Accommodation fails

Fig. 5.9 Details of simulation results

to configure a suitable VN in most cases when the identification of traffic situations failed. When we applied only the BAM-based method, it was necessary to use the attractor-selection-based method in $81.6(= 100 - 18.4)\%$ of trials. By incorporating the linear-regression-based method into our VN reconfiguration framework, we need the attractor-selection-based method only in $15.9(= 12.5 + 3.4)\%$ of trials. This indicates that we can effectively reduce the VN reconfiguration necessary to achieve a suitable VN for the current traffic situation.

5.6 Preparing/Updating Attractors in Yuragi Learning

5.6.1 Approach for Preparing Attractors

As mentioned in Sect. 5.2.3.3, one of the tasks in Yuragi control is to design and prepare attractors in dynamic systems. In this section, we present approaches for preparing and updating attractors in Yuragi learning. A fundamental approach is to deploy our linear-regression-based method in Sect. 5.5.2. That is, we prepare attractors such that we can obtain better VNs when the linear sum of the predefined traffic situations of the attractors expresses a wider variety of traffic situations and decreases the residuals. Therefore, D traffic situations $\boldsymbol{\mu}_1, \cdots, \boldsymbol{\mu}_D$ should be selected so that the linear sum of these traffic situations improves the ability to express various traffic situations.

One way to improve the ability to express various traffic situations is to select a set of attractors so that they have linear independence when each traffic situation is considered as a vector. This is because the dimension of the space spanned by vectors of attractors reaches a maximum when the set of selected traffic situation vectors has linear independence. Whether the set of selected traffic situation vectors has linear independence can be easily judged by checking whether the rank of the matrix $M = [\boldsymbol{\mu}_1, \cdots, \boldsymbol{\mu}_D]$ matches D.

Here, we evaluate the relationship between the ability to express various traffic situations and the performance of the linear-regression-based VN reconfiguration method. Although the evaluation parameters are the same as described in Sect. 5.5.2, we use three sets of attractors as follows:

- Set 1: $\{\boldsymbol{\mu}_1, \boldsymbol{\mu}_2, \boldsymbol{\mu}_3, \boldsymbol{\mu}_4, \boldsymbol{\mu}_5\}$,
- Set 2: $\{\boldsymbol{\mu}_1, \boldsymbol{\mu}_2, \boldsymbol{\mu}_3, \boldsymbol{\mu}_4, (\boldsymbol{\mu}_1 + \boldsymbol{\mu}_2)/2\}$,
- Set 3: $\{\boldsymbol{\mu}_1, \boldsymbol{\mu}_2, \boldsymbol{\mu}_3, (\boldsymbol{\mu}_1 + \boldsymbol{\mu}_2)/2, (\boldsymbol{\mu}_2 + \boldsymbol{\mu}_3)/2\}$.

The number of attractors in each set is five. The rank of matrix M with each traffic situation as a column is 5 for the first set, 4 for the second set, and 3 for the third set. That is, the first set has linear independence, whereas the second and third sets are linearly dependent. The first set is the same as that used for the evaluation in Sect. 5.5.2.

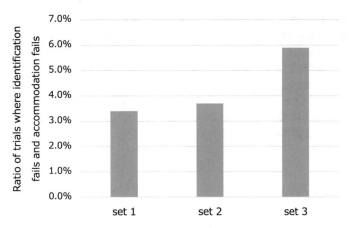

Fig. 5.10 Effect of linear independence of sets of attractors

Figure 5.10 presents the ratio of trials in which the identification of traffic situations fails and the VN configured by the linear-regression-based method cannot accommodate the traffic demand out of 1000 trials, which we call the failure rate, when using each set of attractors. Figure 5.10 indicates that the failure rate is reduced by using the first set of traffic situations. Moreover, it can be seen that the failure rate is smaller in the case of using the second set with rank 4 than in the case of using the third set with rank 3. In other words, the failure rate decreases as the rank of matrix M with each selected traffic situation as a column increases.

The above evaluation indicates that it is effective to maximize the rank of matrix M with each selected traffic situation as a column; that is, it is effective to select a set of attractors so that is has linear independence.

5.6.2 Approach for Updating Attractors

Our VN reconfiguration framework can be applied to dynamically update the set of attractors to accommodate changing traffic demand in the future. When the identification of traffic situations fails and our linear-regression-based method can configure a VN suitable for the current traffic situation, it is not necessary to update the set of attractors, as our VN reconfiguration framework can handle traffic fluctuations using the stored traffic situations. In the case in which the VN configured by the linear-regression-based method cannot accommodate the traffic demand, we apply the attractor-selection-based method to identify a suitable VN. When a solution (i.e., a VN that can accommodate the traffic demand) is identified, we add the traffic situation at that time to the set of attractors. However, because it takes a longer time to identify traffic situations as the number of attractors increases, as discussed in Sect. 5.4.4, it is not desirable for the number of attractors to be too large. Therefore, it is necessary to delete some of the attractors from the set. We believe that it is sufficient to delete traffic situations from the set with a longer time period rather than add new traffic situations. We can therefore use information about the current traffic demand matrix $\mathbf{T}_{now}$ when we delete some traffic situations from the set. We denote the corresponding traffic situation μ_{now} when the traffic demand matrix is $\mathbf{T}_{now}$. One guideline to delete some of traffic situations from the set is to select a traffic situation that can be represented by a linear sum of the predefined traffic situations of the attractors.

5.7   Conclusion

In this chapter, we introduced the concept of Yuragi learning, which deploys a model of human perceptual decision-making. Yuragi learning controls the state of network systems by predefined attractors by the observation and accumulation of sensory information over a period of time. As one application of Yuragi learning, we described virtual network control, which changes the virtual connectivity of nodes in accordance with changes in the traffic demand. Numerical examples demonstrated that Yuragi learning reconfigures the virtual network to the most suitable candidate. As a result, the attractor-selection-based method requires only little effort to identify a suitable VN, although it requires several observations to identify the traffic situation.

References

1. Bitzer, S., Park, H., Blankenburg, F., Kiebel, S.J.: Perceptual decision making: drift-diffusion model is equivalent to a Bayesian model. Front. Hum. Neurosci. **8**, 102 (2014)
2. Bitzer, S., Bruineberg, J., Kiebel, S.J.: A Bayesian attractor model for perceptual decision making. PLoS Comput. Biol. **11**(8), e1004442 (2015)
3. Christodoulopoulos, K., Tomkos, I., Varvarigos, E.: Elastic bandwidth allocation in flexible OFDM-based optical networks. J. Lightwave Technol. **29**(9), 1354–1366 (2011)
4. Fard, P.R., Park, H., Warkentin, A., Kiebel, S.J., Bitzer, S.: A Bayesian reformulation of the extended drift-diffusion model in perceptual decision making. Front. Comput. Neurosci. **11**, 29 (2017)
5. Haykin, S.S., et al.: Kalman Filtering and Neural Networks. Wiley Online Library, Hoboken (2001)
6. Heekeren, H.R., Marrett, S., Ungerleider, L.G.: The neural systems that mediate human perceptual decision making. Nat. Rev. Neurosci. **9**(6), 467–479 (2008)
7. Hopfield, J.J.: Neurons with graded response have collective computational properties like those of two-state neurons. Proc. Natl. Acad. Sci. **81**(10), 3088–3092 (1984)
8. Knuth, D.: The Art of Computer Programming, vol. 2. Seminumerical Algorithms. Addison-Wesley, Boston (1969)
9. Nucci, A., Sridharan, A., Taft, N.: The problem of synthetically generating IP traffic matrices: initial recommendations. ACM SIGCOMM Comput. Commun. Rev. **35**(3), 19–32 (2005)

10. Seber, G.A., Lee, A.J.: Linear Regression Analysis, vol. 936. John Wiley & Sons, Hoboken (2012)
11. Wang, R., Mukherjee, B.: Spectrum management in heterogeneous bandwidth networks. In: Proceedings of IEEE GLOBECOM, pp. 2907–2911 (2012)

Chapter 6
Fast/Slow-Pathway Bayesian Attractor Model for IoT Networks Based on Software-Defined Networking with Virtual Network Slicing

Onur Alparslan and Shin'ichi Arakawa

Abstract Due to power and space considerations, the link capacity in many Internet of Things (IoT) networks is low; however, many IoT sensors, such as high-resolution video cameras, generate huge amounts of data, which can cause congestion unless traffic engineering is applied. However, the majority of the existing network traffic engineering methods require traffic matrix information, which can be difficult to estimate in IoT networks. Instead of attempting to estimate the traffic matrix, we identify the traffic pattern using machine learning based on the Bayesian attractor model (BAM) for supervision and automation of traffic engineering in IoT networks that exhibit a limited number of traffic patterns. We propose running two BAMs in parallel: a fast-pathway BAM for fast but low-certainty identification, and a slow-pathway BAM for slow but high-certainty identification. We demonstrate that our framework enables fast and reliable identification of traffic patterns. After identifying a traffic pattern, a network configuration that is optimized for the identified pattern by traffic engineering is applied to minimize the maximum link utilization. In traffic engineering, we apply virtual network slicing, which creates an independent end-to-end logical network for each IoT sensor type on a shared physical infrastructure. We demonstrate that virtual network slicing allows for fine-grained traffic engineering in IoT networks.

6.1 Introduction

In many networks, several management tasks, such as traffic engineering, are generally performed with human intervention [17]. However, the complexity of networks is increasing with the continuous introduction of new networking technologies, such as new hardware and protocols. Moreover, the number of nodes and routing paths

O. Alparslan (✉) · S. Arakawa
Graduate School of Information Science and Technology, Osaka University, Suita, Osaka, Japan
e-mail: a-onur@ist.osaka-u.ac.jp; arakawa@ist.osaka-u.ac.jp

© Springer Nature Singapore Pte Ltd. 2021
M. Murata, K. Leibnitz (eds.), *Fluctuation-Induced Network Control and Learning*,
https://doi.org/10.1007/978-981-33-4976-6_6

are rising exponentially with the increasing popularity of the Internet of Things (IoT). This rising complexity makes it increasingly difficult to manage networks by traditional human intervention.

Software-defined networking (SDN) is a centralized approach to network management that uses software-based controllers with application interfaces. SDN simplifies network monitoring, configuration, and provisioning. Traditionally, network devices are composed of mainly a control plane and data plane. When a new packet arrives, the control plane determines the destination port to forward the packet. Then, the data plane handles the processing, modification, and forwarding of the packet. SDN decouples the data and control planes and standardizes the control plane while centralizing the control, which enables configuration of the data plane of many networking devices at the same time by standards-based and vendor-neutral protocols. The centralized SDN controller determines and establishes the packet forwarding rules on the network devices. Furthermore, SDN allows for the administration of the SDN controller and the addition of new services using SDN applications via northbound application programming interfaces (APIs). In the past, network administrators were usually required to purchase proprietary hardware or software from the vendor of the existing hardware in the network to add new network services. However, the open API of SDN allows third parties to develop applications that can add services and replace or extend the existing services implemented in the firmware of network devices. Administrators can select the software best suited to their needs from a wide range of implementations by multiple vendors without vendor lock-in; as a result, they are not constrained to one vendor's proprietary limitations.

Traditional traffic engineering methods require network-wide traffic matrix information for optimizing link utilization to prevent congestion. Although SDN enables network operators to collect some statistics from multiple network devices, it is difficult to estimate the traffic matrix in complex networks even with the network monitoring functions of SDN. There have been a number of proposals in the literature for estimating the traffic matrix in SDN networks; however, most use flow-level statistics for the estimation, which have various trade-offs [6, 10, 11, 18]. Furthermore, although SDN provides useful tools for managing complex networks, it requires external supervision. However, the rising complexity of networks makes it increasingly difficult to achieve this supervision by human intervention or a precalculated set of rules. Nevertheless, SDN allows for the development and addition of new services via APIs without vendor lock-in; therefore, many novel applications for solving these problems have been introduced by third-party developers. Among these solutions, machine learning (ML)-based solutions have become popular, as they have advantages over traditional approaches with respect to the supervision and automation of networks because they can identify anomalies in the network, project traffic trends, and make smart decisions without human intervention [9, 21]. Moreover, SDN can further improve the efficiency of ML algorithms; thus, there is mutual benefit. ML applications can receive statistics from all devices in the network via SDN and can therefore provide more accurate decisions compared to running an ML application on a standalone network device with statistics from that device alone. Moreover, SDN allows for the application of the network configuration determined by the ML algorithm to all network devices in real time. SDN also provides a northbound API for applications, which makes it possible to run the ML algorithm on specialized external hardware instead of network devices to reduce the cost and complexity of network devices.

IoT is a network architecture that can greatly benefit from the combined use of SDN, ML, and network slicing. Advances in electronics, which have decreased the power consumption, size, and cost of devices, have led to an exponential explosion of IoT devices and real-time data on the Internet generated by these devices. However, many IoT networks have low link capacity, and some IoT devices generate traffic without congestion control due to power requirements. Nevertheless, many IoT sensors, such as high-resolution cameras, generate huge amounts of data that can cause congestion in IoT networks. Traffic engineering methods can be applied to solve the congestion; however, most methods in the literature require up-to-date traffic matrix information, which can be difficult to estimate with traditional flow-based methods.

Some IoT networks, such as some surveillance systems, are only composed of sensors that produce traffic with highly discrete patterns. For example, many audio and video sensors apply constant bitrate (CBR) or average bitrate (ABR) encoding. The sensors can choose a bitrate from a list depending on the time or the environment conditions. Because the traffic sources have a limited number of mean bitrate options, the IoT networks with such sensors exhibit a limited number of mean traffic matrices, which can be determined before running the network. In [1], we proposed an SDN framework that uses the brain-inspired Bayesian attractor model (BAM) [3] to identify the current traffic matrix in IoT networks that exhibit a limited number of traffic patterns. The BAM is an ML algorithm that models the brain's cognitive abilities and can learn from previous data and identify patterns in real-time noisy data samples. Previously, we applied the BAM to identify a virtual network suitable for traffic in optical networks [15]. Later, we adapted the BAM to SDN and IoT and applied the BAM to identify the current traffic matrix from the statistics of link utilization [1]. An identification can be simpler with less bandwidth and processing requirements than calculating a traffic matrix. In [1], we presented a detailed comparison of our method and the methods in the literature. Moreover, we demonstrated that identification can be performed using a limited set of data (e.g., edge link statistics instead of all links), which can greatly reduce the amount of data required for identification.

In this work, we demonstrate that using two BAMs in parallel (i.e., slow-pathway BAM and fast-pathway BAM, as in the brain) further increases the speed of pattern identification when the number of pattern candidates is high. After a new traffic pattern is identified, our framework updates the routing table in the network according to a traffic engineering algorithm to minimize the maximum link utilization to prevent or alleviate congestion. Although we use the congestion level as the main QoS metric in this work, different traffic engineering methods can be applied to optimize other QoS metrics, such as the round-trip delay time and packet drop ratio. Because the traffic matrices that the network may exhibit are known, optimum routing tables can be pre-calculated by the traffic engineering algorithm and applied by SDN as soon as a new traffic matrix is identified. We also incorporate virtual network slicing by SDN in our framework, which can greatly improve the QoS and security in heterogeneous IoT networks by applying slice-specific traffic engineering and policies [4, 8]. The centralization of the control plane by SDN enables the dynamic creation and control of virtual networks by network slicing [7]. Simulations reveal that our framework can identify a changing traffic pattern by the BAM and optimize the network for the new pattern by applying a network configuration with SDN that is optimized by traffic engineering with network slicing. In this work, we demonstrate our framework on an IoT network; however, it can also be applied to other types of networks that exhibit patterns in their traffic matrices.

The remainder of this chapter is organized as follows. Section 6.2 introduces the BAM, Sect. 6.3 explains the architecture of our framework, Sect. 6.4 describes the simulation scenario and presents the simulation results, and Sect. 6.5 concludes the paper.

6.2 Bayesian Attractor Model

In this section, we describe the decision-making process of the brain. Then, we describe the BAM, which models the decision model of the brain analytically. Finally, we explain the slow-pathway/fast-pathway BAM model, which further increases the identification speed of the BAM when there are many attractors.

6.2.1 Decision-Making Process of the Brain

It is widely accepted that the brain continuously collects noisy sensory information from the surroundings using sensory organs and makes decisions using this information after enough information is accumulated. Even if there is not enough sensory information to make a reliable decision, partial information can be sufficient to make fast and coarse decisions in urgent situations, which can be a matter of life or death. Studies have revealed that there are two pathways in the brain that work in parallel for making decisions: a fast pathway and a slow pathway [2, 19]. The fast pathway produces fast but coarse decisions on important events with low perceptual evidence. For example, when a long thin object suddenly appears in a forest, the fast pathway can identify it as a possible snake and rapidly alert the individual to take precautions to prevent a snake bite. In contrast, the slow pathway takes more time to generate a decision; however, its decisions are more reliable. After seeing a long thin object and taking precautions against a possible snake, the brain can obtain additional perceptual evidence from sensory organs for the slow pathway by observing the tail and head of the object and viewing it from different angles to determine whether it is a snake, rope, or something else. Even if the object is not a snake and the decision by the fast pathway to take precautions is incorrect, the slow pathway corrects this decision later. Taking unnecessary precautions for a short time until the slow pathway decides that they are unnecessary generates less harm than the damage of being bitten by a snake. If the decision by the fast pathway is correct, the precautions can prevent the individual from being bitten. Therefore, the parallel usage of fast and slow pathways in the brain can greatly increase the probability of survival.

6.2.2 The Analytical Model of BAM

For the identification of traffic patterns from noisy link utilization statistics, we use the BAM that models the decision-making process of the brain [3]. The BAM combines the concept of the Bayesian inference model and the attractor model. In the pure attractor model, a state variable is assigned to each decision, and the value of the state variable indicates how much evidence is accumulated to support the corresponding decision. The values of the state variables are continuously pulled by an attractor while being reduced by the value of other state variables. The pulling force between an attractor and a state variable increases with the distance between them so that the state variable does not drift away from the attractor. As the value of a state variable increases, the state variable applies a higher force reducing the value of other attractors so that the winning decision becomes clearer based on the gap between the values of the state variables.

Noisy sensory information enters the system as an external impulse and drives the attractor dynamics, and the state variables accumulate evidence from the external impulses. The state variable that reaches the threshold is selected as the decision. The drawback of the pure attractor model is that it does not consider the properties of noisy sensory information (e.g., variance, noise level), which may impact the results, because there is no feedback from the decision state to the sensory system. However, recent experiments have revealed that there is a feedback mechanism in the brain from ongoing decisions to the sensory system, which modulates the impact of sensory information [3]. The BAM models this feedback by applying Bayesian inference with a generative model.

Bayesian inference-based approaches are widely used in modeling the brain's cognitive abilities and human behaviors when making decisions. Bayesian inference combines noisy sensory evidence with the internal dynamics of decision-making. The generative model in the Bayesian inference of the BAM predicts a probability distribution over the observations based on its attractor dynamics and the current decision state. This probability distribution of observations is used as feedback to the sensory level to modulate the extraction of evidence from sensory information.

The BAM assumes that the brain assigns stochastic variables to possible decisions (attractors). The BAM continuously updates the posterior probability of these stochastic variables using evidence from the observations and the attractor dynamics. This part of the model is called the cognitive process. After the confidence of a decision (attractor) becomes sufficiently high, the BAM selects this attractor as its decision. This part of the model is called the decision-making process.

The analytical model of the BAM is as follows. The BAM has several fixed points ϕ_i, each of which corresponds to a decision alternative (attractor) i. The value of attractor i is vector μ_i. For each attractor, there is a decision state variable z_i, which is updated according to an internal generative model. The state variables are stored in vector $\mathbf{z}$. When a new observation is made at time t, the BAM calculates the posterior distribution of the state variable $\mathbf{z}_t$ by $p(\mathbf{z}_t \mid \mathbf{X}_{\Delta t:t})$ using the observations denoted by $\mathbf{X}_{\Delta t:t}$, where t is the current time and Δt is the step size of the observation time.

Unlike the pure attractor model, where there is no feedback from the decision state to the sensory level, the decision state in the BAM affects both the internal predictions and the gain of the sensory evidence. The generative model in the BAM models the change in the decision state according to Hopfield dynamics. It updates the decision state $\mathbf{z}$ from one step to the next step at time t by the equation

$$\mathbf{z}_t - \mathbf{z}_{t-\Delta t} = \Delta t f(\mathbf{z}_{t-\Delta t}) + \sqrt{\Delta t}\,\mathbf{w}_t, \tag{6.1}$$

where $f(\mathbf{z})$ is the function of the Hopfield dynamics, and $\mathbf{w}_t$ is a Gaussian noise variable with distribution $\mathbf{w}_t \sim \mathcal{N}(\mathbf{O}, \mathbf{Q})$, where $\mathbf{Q} = (q^2/\Delta t)\mathbf{I}$ is the covariance of the noise process, and q is a parameter representing dynamics uncertainty. $\mathbf{w}_t$ is the expected state noise at the attractor level showing a propensity to switch between attractors. Higher dynamics uncertainty signifies that it is more likely for state switches to occur between decision alternatives. This part of the algorithm is the first step in the BAM, as illustrated in Fig. 6.1.

Given a decision state $\mathbf{z}$, [3] predicts the probability distribution over observations using the attractors by the generative model

$$\mathbf{x} = \mathbf{M}\sigma(\mathbf{z}) + \mathbf{v}, \tag{6.2}$$

Fig. 6.1 Illustration of main steps of Bayesian attractor model

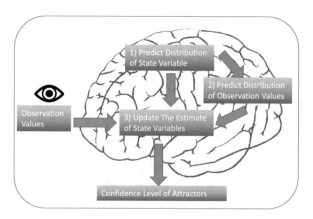

where $\mathbf{M} = [\boldsymbol{\mu}_1, \cdots, \boldsymbol{\mu}_N]$ is the set of the attractors (mean feature vectors) and $\boldsymbol{\sigma}$ is the sigmoid function, which maps all state variables z_i to values between 0 and 1. The noise variable $\mathbf{v}$ has a normal distribution $\mathcal{N}(\mathbf{0}, \mathbf{R})$, where $\mathbf{R} = r^2\mathbf{I}$ is the expected isotropic covariance of the noise in the observations and r is the sensory uncertainty. A higher r signifies that higher noise is expected in the observations. This part of the algorithm is the second step in the BAM, as illustrated in Fig. 6.1.

When the state variable z_i is low, the sigmoid function produces a result close to 0. However, its result is still a positive number. If there are many attractors, the cumulative sum of the sigmoid multipliers of low state values can reach a significantly high value and may even exceed 1, while the maximum sigmoid multiplier of an attractor with a high state value is limited to 1. The presence of many attractors may cause a high deviation in the generative function in Eq. (6.2) and cause identification failure. In this work, we use many attractors in the slow-pathway BAM and thus apply a different equation,

$$\mathbf{x} = \frac{\mathbf{M}\sigma(\mathbf{z})}{\Sigma\sigma(z_i)} + \mathbf{v}, \tag{6.3}$$

as the generative function of the slow-pathway BAM, where $\Sigma\sigma(z_i)$ sums the sigma functions of the state variables to normalize the generative function.

The BAM infers the distribution of the decision state, which is $p(\mathbf{z}_t \mid \mathbf{X}_{\Delta t:t-\Delta t}) \approx \mathcal{N}(\hat{\mathbf{z}}_t, \hat{\mathbf{P}}_t)$, using the unscented Kalman filter (UKF) [20]. Although other filters, such as the extended Kalman filter, can also be used, the UKF is selected as it has a reasonable trade-off between the approximation deviation and the computation time. Consequently, the UKF is used to approximate the distribution of observations, which is $p(\mathbf{x}_t \mid \mathbf{X}_{\Delta t:t-\Delta t}) \approx \mathcal{N}(\hat{\mathbf{x}}_t, \hat{\boldsymbol{\Sigma}}_t)$, in the next step. The prediction error between the predicted mean $\hat{\mathbf{x}}_t$ and the actual observation $\mathbf{x}_t$ is

$$\epsilon_t = \mathbf{x}_t - \hat{\mathbf{x}}_t. \tag{6.4}$$

Then, the decision state prediction $\hat{\mathbf{z}}_t$ and its posterior covariance matrix $\bar{\mathbf{P}}_t$ are updated via Kalman gain $\mathbf{K}_t$:

$$\bar{\mathbf{z}}_t = \hat{\mathbf{z}}_t + \mathbf{K}_t\epsilon_t, \tag{6.5}$$

$$\bar{\mathbf{P}}_t = \hat{\mathbf{P}}_t - \mathbf{K}_t\hat{\mathbf{C}}_t^T. \tag{6.6}$$

The Kalman gain is calculated by

$$\mathbf{K}_t = \hat{\mathbf{C}}_t\hat{\boldsymbol{\Sigma}}_t^{-1}, \tag{6.7}$$

where $\hat{\mathbf{C}}_t$ is the cross-covariance between the predicted decision state $\hat{\mathbf{z}}_t$ and the predicted observation $\hat{\mathbf{x}}_t$, and $\hat{\boldsymbol{\Sigma}}_t$ is the covariance matrix of the predicted observations. This part of the algorithm is the third step in the BAM, as illustrated in Fig. 6.1.

In [3], the value of $p(\mathbf{z}_t = \phi_i \mid \mathbf{X}_{\Delta t:t})$, which is the posterior density over the decision state evaluated at the stable fixed point ϕ_i, is used as a confidence level metric of each attractor i. Although it is possible to use the posterior density calculation as a confidence level metric, we instead use the state value z_i in this work, as z_i is less oscillatory and more clearly bounded than the posterior density $p(\mathbf{z}_t = \phi_i \mid \mathbf{X}_{\Delta t:t})$.

For the decision criterion, we use the difference between the state values z_i of attractors. The state values are sorted from highest to lowest. Let attractor j have the highest state value and attractor k have the second highest state value. If the difference between the state values of these two attractors is higher than threshold λ, then attractor j is identified as the new state. That is,

$$z_j - z_k > \lambda. \tag{6.8}$$

The fast-pathway BAM uses λ_f, while the slow-pathway BAM uses λ_s as the decision threshold. Selecting a high λ can increase the reliability of identification; however, the identification may take a longer time.

6.2.3 Fast/Slow-Pathway Bayesian Attractor Model

As the number of attractors stored in the BAM increases, the convergence speed of the state variables may decrease, and the identification may take more time. As described in Sect. 6.2.1, for faster identification of important situations, the brain uses two pathways: the fast pathway that makes fast but coarse decisions and the slow pathway that makes slower but more reliable decisions. Imitating the brain, we use a fast-pathway BAM and slow-pathway BAM concurrently in our framework for faster identification. Figure 6.2 illustrates the fast/slow-pathway decision scheme in our framework. When new observation data arrive, the data are sent to the fast-pathway BAM and slow-pathway BAM at the same time. The difference between the fast-pathway BAM and the slow-pathway BAM is that whereas the slow-pathway BAM stores all possible attractors, the fast-pathway BAM stores a smaller set of attractors. Because the fast-pathway BAM has a smaller solution space, its decision states may converge faster than the slow-pathway BAM with similar parameters. In addition, using fewer attractors may make it possible to use a more aggressive parameter set for even faster identification.

The solution space of the fast-pathway BAM can be kept low by using only attractors with the highest priority. Moreover, the processing and convergence speed of the BAM can be further increased by reducing the dimension of the attractor vectors $\boldsymbol{\mu}$. For example, when a suspicious object appears, the shape of the object alone can be sufficient for the brain to classify it as a possible hazard. Namely, identifying the shape of an object as long and thin by the fast pathway in the brain

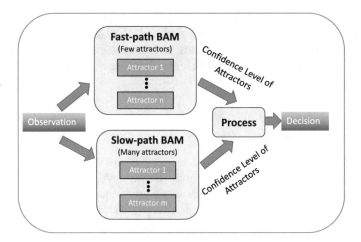

Fig. 6.2 Illustration of fast-pathway/slow-pathway Bayesian attractor model decision scheme

allows an individual to take precautions against a possible snake. Later, the slow pathway can provide a more reliable result by examining other features, such as the presence of teeth, legs, scales, and eyes, to identify the object as a snake or rope. Even if the object is identified as a snake, the slow pathway can verify additional features, such as the color, head, and pupils of the snake, to identify its species and whether it is venomous. Similarly, in a network, the traffic patterns that are more likely to cause congestion can be selected as fast-pathway attractors for faster identification. Moreover, links that are less indicative of whether a traffic pattern may cause congestion can be removed from the attractor vectors of the fast-pathway BAM for faster processing and a higher convergence speed.

As the fast-pathway and slow-pathway BAMs work in parallel, both calculate the confidence levels of their attractors simultaneously and independently. The control algorithm processes the confidence level results of both BAMs and attempts to identify the latest status of the system. The fast-pathway BAM and slow-pathway BAM can produce contradictory results; therefore, the results must be processed carefully. The steps of an iteration of our decision algorithm are as follows:

1. After a new observation arrives, process the observation in both BAMs and obtain their confidence level results.
2. Because the fast-pathway BAM can converge faster, it can provide the most up-to-date result. Therefore, first check the result of the fast-pathway BAM. If there is a new attractor that passes the state value difference threshold λ_f in the fast pathway, identify this attractor as the new status and end the iteration. Otherwise, proceed to the next step.
3. If there is a new attractor that passes the state value difference threshold λ_s in the slow pathway, identify this attractor as the new status and end the iteration. Otherwise, proceed to the next step.

4. If the attractor of the fast-pathway BAM was previously selected but the slow-pathway BAM has identified a different attractor (passing the state value difference threshold λ_s) for a long time, select the attractor of the slow-pathway BAM as the new status, as the reliability of the slow-pathway BAM is higher. End the iteration.

6.3 Proposed Architecture

Most traffic engineering methods require traffic matrix information. However, due to the large number of source–destination pairs and large number of flows, estimating the traffic matrix precisely in a large Internet Protocol (IP) network is extremely difficult and has been recognized as a challenging research problem [22]. Although SDN does not fully solve the problem, it has several important advantages for traffic engineering. First, the SDN controller has a global view of the SDN network topology. Second, the SDN controller can collect flow-level and port (link)-level statistics from SDN switches by using a southbound protocol, such as OpenFlow [12]. However, traffic matrix estimation remains challenging even with SDN. For example, there are hybrid networks that contain non-SDN-capable switches, and it is not possible to obtain traffic statistics from non-SDN switches by OpenFlow.

Another problem is that OpenFlow supports collecting only flow- and link-level statistics without direct support to obtain source–destination-based traffic statistics. To estimate the traffic matrix by SDN, it is necessary to collect the per-flow statistics from all switches. For large-scale networks with many flows, such as a large-scale IoT network, transmitting the statistics of all flows in the network to the controller periodically (e.g., every few seconds) may require high bandwidth, which can be problematic in IoT networks where the link capacity may be low. Moreover, to calculate the overall traffic matrix, periodically processing the statistics of a large number of flows in real time is a central processing unit (CPU)-intensive task. There have been a number of proposals in the literature for traffic matrix estimation by SDN; however, they have a trade-off between the processing overhead, transmission overhead, and accuracy of traffic matrix estimation [6, 10, 11, 18].

Instead of attempting to calculate the overall traffic matrix in IoT networks that exhibit a limited number of traffic patterns, our BAM-based traffic engineering framework aims to identify the traffic pattern using only the utilization level information of a set of edge links. Our method has a number of advantages. First, an identification can be simpler with less bandwidth and processing requirements than calculating a traffic matrix. Moreover, the identification is a direct solution. If the identification is correct, the identification gives the exact mean traffic matrix.

Another problem in traditional traffic engineering is that it is difficult to satisfy application-level QoS requirements. Concurrent IoT applications may have different QoS requirements that may be difficult to satisfy on the same network. For example, in an IoT network of autonomous vehicles, if the vehicles' high-definition video sensors and control systems are on the same network, congestion in the network due to video transmission may disrupt control communication and result in an accident [8]. Network slicing by SDN can solve this problem by creating separate virtual networks over a shared physical network. By logically separating the networks, network slicing can allow IoT services with diverse requirements to exist on the same physical infrastructure, while satisfying their QoS requirements. Moreover, network slicing allows for the isolation of IoT device communications for security [4]. Our framework supports network slicing through SDN and can therefore improve the QoS and security of IoT networks.

The architecture of our framework is as follows. Traffic identification is performed in an SDN application using the BAM. Initially, a list of possible traffic patterns is provided to our application as Bayesian attractors. The application runs fast-pathway and slow-pathway BAMs simultaneously. The slow-pathway BAM uses all attractors, while the fast-pathway BAM uses only a small set of important attractors. Together with the traffic patterns, the network configurations (precalculated network slice topologies and routing tables) optimized for each traffic pattern are stored by our application. The SDN controller collects information, such as the network architecture and link utilization statistics, from the network, and our SDN application periodically receives the utilization statistics of a set of links from the SDN controller. The fast/slow-pathway BAMs update the confidence level of their attractors whenever there is a new observation of link utilization levels. By comparing the confidence levels calculated by both fast/slow-pathway BAMs, the application attempts to identify the current traffic pattern. If the identified traffic pattern changes, the application sends the network configuration, including the routing tables and virtual network slicing information, which has been preoptimized for the new identified traffic, to the SDN controller. The SDN controller then reconfigures the switches according to the new network configuration.

6.4   Simulation Results

To evaluate the performance of our framework, we implemented a network simulator in Matlab and simulated an SDN IoT network. We demonstrated that after an abrupt change in the traffic matrix, our framework was able to identify the new traffic pattern and minimize the maximum link utilization by applying a network configuration optimized for the new pattern. Recently, IoT-based crowd surveillance systems have become popular in many applications, such as public safety (against terrorist attacks, protests, natural disasters, etc.) and smart cities [5, 13, 14, 16]. We simulated a crowd surveillance IoT network with nine core nodes and six edge nodes, where each edge node had two types of sensors. The IoT network topology and the initial traffic rate of the sensors are presented in Fig. 6.3. In the network, the core nodes were SDN-capable, while the edge nodes were not.

In the simulation scenario, a crowd gathered around a building. The sensors of the crowd surveillance system were placed at six locations around the building,

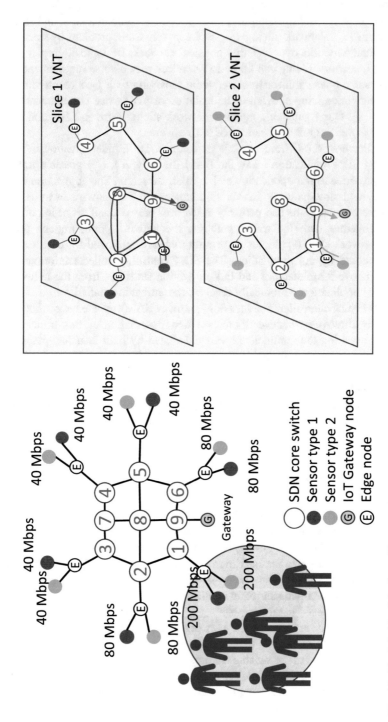

Fig. 6.3 The left side of the figure illustrates the Internet of Things (IoT) network topology and the initial traffic rate of the sensors when the crowd was close to the edge of core node 1. The right side of figure illustrates the virtual network slices optimized for this traffic

and the amount of data generated by a sensor changed with the presence of people in the vicinity of the sensor. Each sensor type generated a different type of data (i.e., infrared video and 360° panoramic video) and sent one-way traffic to an IoT gateway node that forwarded the data generated in the IoT domain to a wide-area network (WAN). For simplicity, we set the mean data generation rate of both sensor types to be equal to each other in the simulation. Whereas the mean data generation rate of the sensors on an edge node was the same, the traffic rate of each sensor changed according to an independent log-normal distribution; thus, their transient rates were different. Each of the sensors closest to the center of the crowd, such as the sensors connected to the edge node of core node 1 in Fig. 6.3, generated 200 Mbps traffic using an ABR video codec with a high bitrate to identify fast-paced movement of many people. The sensors that were in the vicinity but not so close to the center of the crowd, such as the sensors connected to the edge node of core nodes 2 and 6 in Fig. 6.3, generated 80 Mbps ABR traffic. The sensors that were far away from the crowd did not identify much movement so they generated 40 Mbps ABR traffic.

Because the traffic changed based on the presence of a crowd near the sensors of an edge node and there were six edge nodes, the total number of possible traffic patterns was $2^6 = 64$. The attractors corresponded to the mean traffic rate on the edge links after applying the traffic patterns. For example, the BAM attractor of the traffic pattern illustrated in Fig. 6.3 (when the crowd was only at the edge of core node 1) was simply an array [400 160 80 80 80 160] in terms of Mbps, where the array index is the node ID of the core node to which the edge link was connected. Because the traffic entered the SDN network through six core routers and the IoT traffic in the experiment was one way, the size of each BAM attractor was 1×6.

The slow-pathway BAM contained all possible attractors; thus, 64 attractors were stored in the slow-pathway BAM. Among these attractors, three were selected and stored as attractors in the fast-pathway BAM. The attractors of the fast-pathway BAM can be selected according to different objectives. For example, in a network in which the traffic pattern changes frequently, using the most common traffic patterns as attractors in the fast-pathway BAM can increase the adaptation speed of the network by identifying most patterns and changing the configuration in a short time. However, in another network, some traffic patterns may cause high congestion in the network unless a specially optimized network configuration is applied. Using these traffic patterns as attractors in the fast-pathway BAM can decrease the duration of congestion after the traffic changes to one of these patterns. In this simulation, as the attractors of the fast-pathway BAM, we selected three traffic patterns that may cause high congestion unless a specific network configuration is applied. The first attractor represents the case in which the crowd is only around the edge of core node 1, while the second attractor represents the case in which the crowd is only around the edge of core node 3. The third attractor represents the case in which the crowd is only around the edge of core node 6.

We applied virtual network slicing for fine-grained traffic engineering in the IoT network, as described in Sect. 6.3. For each traffic pattern, a network configuration consisting of the topology and routing table of the virtual network slices optimized for the pattern was pre-computed by traffic engineering and stored in our application. For example, a traffic pattern and the topology of network slices optimized for that pattern are illustrated in Fig. 5.3. Each sensor type and its corresponding traffic was assigned to a different slice; therefore, the traffic of sensor type 1 was assigned to network slice 1, and the traffic of sensor type 2 was assigned to network slice 2. Each slice had a dedicated routing table controlled by SDN. Slicing makes it possible to carry the traffic of different sensors on different paths for fine-grained traffic engineering even if the source and destination nodes are the same.

When the network configuration was optimized for a traffic pattern, the objective was to minimize the maximum link utilization and maximize the power efficiency. The optimization was performed in two steps. In the first step, for a selected traffic pattern, the set of routing tables in each network slice that minimized the maximum link utilization in the physical network was calculated. Then, among the solutions produced in the first step, a solution that used the minimum number of physical links when the virtual network slices were superimposed was selected. Disabling links that were unused in all slices also reduced the power consumption. For example, the four physical links in Fig. 6.3 were set to idle because they were not used in any slices. While the virtual network topology in both slices was the same in Fig. 6.3, the slices had different routing tables to minimize the overall maximum link utilization of the physical network. For example, the traffic from the edge of core node 1 to the IoT gateway followed the path $1 \rightarrow 2 \rightarrow 8 \rightarrow 9$ on slice 1, while the traffic between the same nodes followed the path $1 \rightarrow 9$ on slice 2. Although the traffic optimization algorithm produced the same topology for the slices in Fig. 6.3, the topology may differ depending on the placement of sensors and the traffic matrix on each slice.

In the simulation, we tested a scenario in which the crowd was initially only near the edge sensors of core node 1, as in Fig. 6.3. After 20 iterations, the crowd suddenly moved to the edge sensors of core node 5, as illustrated in Fig. 6.4, and caused an abrupt change in the traffic matrix. The sensors generated bursty traffic with a log-normal traffic rate distribution. The traffic rate of the sensors on virtual network slice 1 and slice 2 is presented in Fig. 6.5, where the x-axis represents the iteration number. At each iteration, a new set of traffic statistics was received from the network and used as input to the BAM to identify the current traffic pattern. Initially, the sensors connected to the edge of core node 1 produced the highest traffic with a mean rate illustrated Fig. 6.3 until iteration 20. We refer to this part of the simulation as the first epoch. However, after the crowd moved closer to the edge of core node 5 at iteration 21, the sensors connected to the edge of core node 5 took the lead and generated traffic with a mean rate illustrated in Fig. 6.4 until iteration 40. We refer to this part of the simulation as the second epoch. Finally, the crowd moved back near the edge of core node 1 at iteration 41, and the traffic rate of the sensors connected to core node 1 took the lead again. The sensors generated traffic with a mean rate illustrated in Fig. 6.4 from iteration 41 to the end of the simulation. We refer to this part of the simulation as the third epoch. The resulting edge link utilization is plotted in the third subplot in Fig. 6.5.

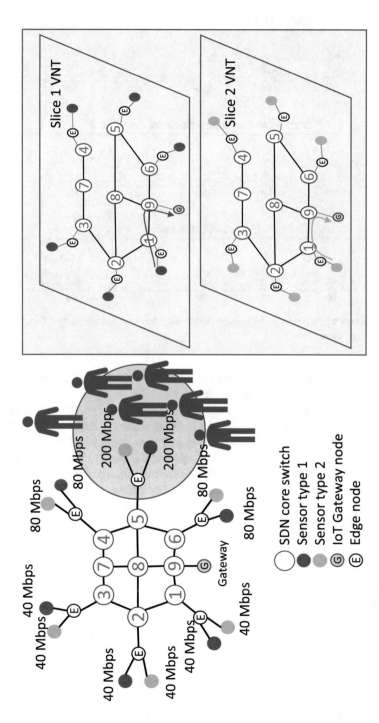

Fig. 6.4 The left side of the figure illustrates the Internet of Things (IoT) network topology and the initial traffic rate of sensors when the crowd was close to the edge of core node 5. The right side of figure illustrates the virtual network slices optimized for this traffic

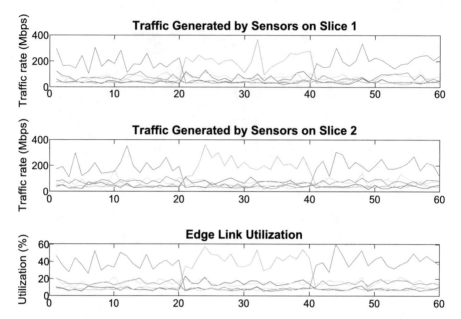

Fig. 6.5 The rate of traffic generated by sensors and the resulting edge link utilization. The x-axis represents the iteration number

The simulation results are presented in Fig. 6.6. The edge link utilization in the third subplot of Fig. 6.5 is replotted in the first subplot of Fig. 6.6 for a clearer presentation of how the simulation results changed with the edge link utilization. The utilization of the links between the core nodes is plotted in the second subplot in Fig. 6.6. At each iteration, the fast-pathway and slow-pathway BAMs received up-to-date edge link utilization as input and updated the state variable z_i of their respective attractors. We used the difference between the values of the state variables as the decision criterion. The state values of the fast-pathway and slow-pathway attractors are plotted in the third and fourth subplots, respectively. Finally, the ID number of the traffic patterns identified by processing the state variables of the fast-pathway and slow-pathway BAMs is plotted in the fifth subplot. The mean traffic matrix between iterations 1 to 20, which is the first epoch, and iterations 41 to 60, which is the third epoch, corresponds the attractor with ID 33. The mean traffic matrix between iterations 21 and 40, which is the second epoch, corresponds to the attractor with ID 3.

In the first epoch, the crowd was near the edge of core node 1; therefore, the edge link of core node 1, which is represented by a blue line in the first subplot of Fig. 6.6, had the highest utilization among all edge links. Because the initial network configuration was set to the configuration optimized for the attractor corresponding to this traffic pattern, which had ID 33, the maximum core link utilization was low in the first epoch, as displayed in the second subplot. Thus, when the network configuration optimized for this traffic pattern is applied, the average utilization of

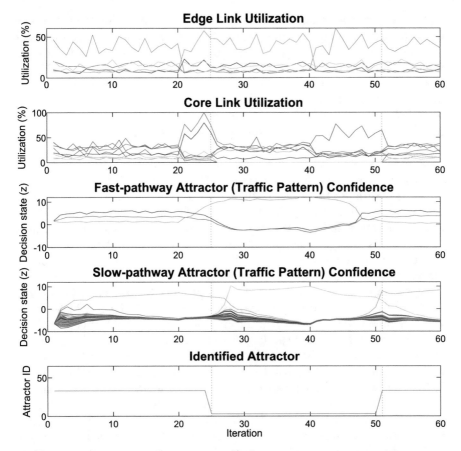

Fig. 6.6 Simulation results. The first subplot displays the utilization of edge links, while the second subplot displays the utilization of core links. The third and fourth subplots display the decision state values of the attractors in the fast-pathway and slow-pathway Bayesian attractor models, respectively. The fifth subplot displays the identified traffic pattern by the decision algorithm

the maximum utilized link is expected to be 32%. The third subplot indicates that the state values of the fast-pathway BAM were close to each other. The attractor corresponding to this traffic pattern was not stored in the fast-pathway BAM, and the fast-pathway BAM was thus unable to identify this traffic. Therefore, the difference between the highest and second highest state variables was less than λ_f, which was set to 4 in the simulation. In contrast, the slow-pathway BAM identified this pattern because the attractor corresponding to this traffic pattern was available in its attractor list. The difference between the green line in the fourth subplot, which represents the state variable of attractor 33, and the second highest state variable exceeded λ_s, which was set to 2 in the simulation; therefore, the traffic pattern was identified properly. Because the fast pathway is more likely to provide incorrect results than the slow pathway, we selected a λ_f higher than λ_s to reduce the rate of incorrect decisions. Because only the slow-pathway BAM identified the pattern and it was the same as the initially selected traffic pattern, the fifth subplot, which represents the identified attractor ID, remained 33 in this epoch.

In the second epoch, the crowd moved to the edge of core node 5; therefore, the traffic changed abruptly at iteration 21, as illustrated in the first subplot. The green line, which represents the utilization of the edge link of core node 5, gained the highest values. However, the virtual topologies and the routing tables of the network slices were far from optimal for this new traffic; thus, the utilization of some of the core links greatly increased, as illustrated in the second subplot. Although the network configuration optimized for the new traffic pattern was able to produce a mean utilization of 32% at the maximum utilized link, the current network configuration, which was optimized for the traffic in the first epoch, produced a mean maximum link utilization of 72%, which is 2.25 times the utilization of the network configuration optimized for this traffic pattern. However, because the attractor corresponding to this traffic pattern (with ID 3) was available in the attractor list of both the fast-pathway and slow-pathway BAMs, both BAMs succeeded in identifying the traffic. The fast-pathway BAM identified the traffic after five iterations when the difference between the highest and second highest state variables passed the threshold λ_f. While the slow-pathway BAM was in the process of identifying the traffic as the traffic pattern of attractor 33, the decision algorithm selected the decision of the fast-pathway BAM as the new traffic pattern, as seen in the fifth subplot, and applied the network configuration optimized for pattern 3. Because the identification was correct, the mean utilization of the maximum utilized core link decreased to 32%. However, the slow-pathway BAM identified new traffic at iteration 28 when the difference between the highest and second highest state values exceeded λ_s. The slow-pathway BAM produced the same result as the fast-pathway BAM, which confirmed that the decision of the fast-pathway BAM at iteration 25 was correct. As a result, the joint usage of fast-pathway and slow-pathway BAMs reduced the identification time by 37% compared to using the slow-pathway BAM alone. The iterations when a new decision was generated are indicated by dotted vertical lines in all subplots.

In the third epoch, the crowd moved back to the edge of core node 1; thus, the traffic changed abruptly at iteration 41, as seen in the first subplot. Again, the network configuration was not optimal for the new traffic pattern; therefore, the mean utilization of the maximum utilized core link increased to 56%, which was 1.75 times the mean utilization of the maximum utilized core link when a network configuration optimized for this traffic was applied. The state variables of both fast-pathway and slow-pathway BAMs started to change after the traffic changed. However, because the new traffic pattern was not available in the attractor list of the fast-pathway BAM, the difference between the highest and second highest state variables could not exceed λ_f. This signifies that the identification was not clear; therefore, the decision algorithm ignored the fast-pathway BAM in this epoch. However, the slow-pathway BAM identified the new traffic pattern after 11 iterations, and the decision algorithm updated the identified pattern and applied the network configuration optimized for this pattern, which reduced the mean utilization of the maximum utilized core link to 32%. Even if the fast-pathway BAM identified an incorrect traffic pattern, the decision algorithm would first apply the result of the fast-pathway BAM and correct it later by switching to the correct result of the slow-pathway BAM after some iterations. This is because the final step of the decision algorithm selects the result of the slow-pathway BAM if the results of the fast-pathway and slow-pathway BAMs differ for a long time.

6.5   Conclusion

Computer networks are becoming increasingly heterogeneous and complex with the deployment of new technologies. This increasing complexity makes it difficult to manage and optimize the networks manually and results in the need for automation and ML techniques. In this chapter, we presented an ML framework based on SDN for optimizing the network configuration in an IoT network by identifying traffic patterns with the joint usage of fast-pathway and slow-pathway BAMs. The simulation results revealed that our framework can identify traffic patterns by using edge link traffic statistics. In addition, our framework applies a network configuration optimized for the identified traffic pattern by SDN. Our framework also uses virtual network slicing, which improves the QoS, security, and power efficiency of IoT networks. The simulation results also revealed that the joint usage of fast-pathway and slow-pathway BAMs improves the identification time compared to using a slow-pathway BAM alone.

In future work, we will implement and test the proposed framework on a real testbed. We will also work on further increasing the traffic identification speed and extending the framework for different types of networks, such as cellular networks.

References

1. Alparslan, O., Arakawa, S., Murata, M.: SDN-based control of IoT network by brain-inspired Bayesian attractor model and network slicing. Appl. Sci. **10**(17), 5773 (2020)
2. Benson, E.: The synaptic self. Monit. Psychol. **33**(10), 40 (2002)
3. Bitzer, S., Bruineberg, J., Kiebel, S.J.: A Bayesian attractor model for perceptual decision making. PLoS Comput. Biol. **11**(8) (2015)
4. Esaki, H., Nakamura, R.: Overlaying and slicing for IoT era based on Internet's end-to-end discipline. In: 2017 IEEE International Symposium on Local and Metropolitan Area Networks (LANMAN), pp. 1–6. IEEE, Piscataway (2017)
5. Gribaudo, M., Iacono, M., Levis, A.H.: An IoT-based monitoring approach for cultural heritage sites: the Matera case. Concurren. Comput. Pract. Exp. **29**(11), e4153 (2017)
6. Hsu, C.Y., Tsai, P.W., Chou, H.Y., Luo, M.Y., Yang, C.S.: A flow-based method to measure traffic statistics in software defined network. In: Proceedings of the Asia-Pacific Advanced Network, vol. 38, p. 19. APAN, Peradeniya (2014)

7. Jain, R., Paul, S.: Network virtualization and software defined networking for cloud computing: a survey. IEEE Commun. Mag. **51**(11), 24–31 (2013)

8. Kafle, V.P., Fukushima, Y., Martinez-Julia, P., Miyazawa, T., Harai, H.: Adaptive virtual network slices for diverse IoT services. IEEE Commun. Stand. Mag. **2**(4), 33–41 (2018)

9. Latah, M., Toker, L.: Artificial intelligence enabled software-defined networking: A comprehensive overview. IET Netw. **8**, 79–99 (2019)

10. Li, M., Chen, C., Hua, C., Guan, X.: Cflow: A learning-based compressive flow statistics collection scheme for SDNs. In: IEEE International Conference on Communications (ICC), pp. 1–6. IEEE, Piscataway (2019)

11. Malboubi, M., Peng, S.M., Sharma, P., Chuah, C.N.: A learning-based measurement framework for traffic matrix inference in software defined networks. Comput. Electr. Eng. **66**(C), 369–387 (2018)

12. McKeown, N., Anderson, T., Balakrishnan, H., Parulkar, G., Peterson, L., Rexford, J., Shenker, S., Turner, J.: OpenFlow: Enabling innovation in campus networks. SIGCOMM Comput. Commun. Rev. **38**(2), 69–74 (2008)

13. Memos, V.A., Psannis, K.E., Ishibashi, Y., Kim, B.G., Gupta, B.: An efficient algorithm for media-based surveillance system (EAMSuS) in IoT smart city framework. Futur. Gener. Comput. Syst. **83**, 619–628 (2018)

14. Motlagh, N.H., Bagaa, M., Taleb, T.: UAV-based IoT platform: a crowd surveillance use case. IEEE Commun. Mag. **55**(2), 128–134 (2017)

15. Ohba, T., Arakawa, S., Murata, M.: Bayesian-based virtual network reconfiguration for dynamic optical networks. IEEE/OSA J. Opt. Commun. Netw. **10**(4), 440–450 (2018)

16. Pathan, A.S.K.: Crowd assisted networking and computing. CRC Press, Boca Raton (2018)

17. Pereira, V., Rocha, M., Cortez, P., Rio, M., Sousa, P.: A framework for robust traffic engineering using evolutionary computation. In: Doyen, G., Waldburger, M., Čeleda, P., Sperotto, A., Stiller, B. (eds.) Emerging Management Mechanisms for the Future Internet, pp. 1–12. Springer, Berlin (2013)

18. Queiroz, W., Capretz, M.A., Dantas, M.: An approach for SDN traffic monitoring based on big data techniques. J. Netw. Comput. Appl. **131**, 28–39 (2019)

19. Soares, S.C., Maior, R.S., Isbell, L.A., Tomaz, C., Nishijo, H.: Fast detector/first responder: Interactions between the superior colliculus-pulvinar pathway and stimuli relevant to primates. Front. Neurosci. **11**, 67 (2017)

20. Wan, E.A., Van Der Merwe, R.: The unscented Kalman filter for nonlinear estimation. In: Proceedings of the IEEE 2000 Adaptive Systems for Signal Processing, Communications, and Control Symposium, pp. 153–158. IEEE, Piscataway (2000)

21. Xie, J., Yu, F.R., Huang, T., Xie, R., Liu, J., Wang, C., Liu, Y.: A survey of machine learning techniques applied to software defined networking (SDN): Research issues and challenges. IEEE Commun. Surv. Tutorials **21**(1), 393–430 (2019)

22. Zhao, Q., Ge, Z., Wang, J., Xu, J.: Robust traffic matrix estimation with imperfect information: making use of multiple data sources. SIGMETRICS Perform. Eval. Rev. **34**(1), 133–144 (2006)

Chapter 7
Application to IoT Network Control: Predictive Network Control Based on Real-World Information

Yuichi Ohsita

Abstract In this chapter, I describe the application of the Yuragi learning mechanism to the control of networks that accommodate Internet of Things (IoT) services. Numerous types of services have been provided through networks as IoT devices have become increasingly popular, and the traffic resulting from such services must be accommodated to satisfy the service-specific requirements. One approach to accommodate traffic is the use of network slicing, which provides multiple network slices for network services. Resources for each slice should be dynamically allocated to follow the traffic changes. Predictive network control is a method for accommodating fluctuating traffic without degrading service quality. This method allocates resources based on predicted future traffic. To predict future traffic, real-world information is useful. Thus, in this chapter, we discuss the use of real-world information related to IoT services to predict future traffic. However, it is difficult to model the relationship between real-world information and future traffic; therefore, we apply Yuragi learning, which is inspired by the cognitive processes of the human brain and makes decisions based on uncertain information. In this chapter, we discuss the application of Yuragi learning to predictive network control based on real-world information.

7.1 Introduction

In recent years, numerous types of devices have become connected to the Internet. Such devices are known as Internet of Things (IoT) devices. As an increasing number of devices have become connected, various services have been provided through the Internet, and networks must accommodate traffic to fulfill the service requirements. The requirements differ for various services: one service may require low-latency communication between devices, whereas another service may require

Y. Ohsita (✉)
Institute for Open and Transdisciplinary Research Initiatives, Osaka University, Suita, Osaka, Japan
e-mail: y-ohsita@ist.osaka-u.ac.jp

© Springer Nature Singapore Pte Ltd. 2021
M. Murata, K. Leibnitz (eds.), *Fluctuation-Induced Network Control and Learning*,
https://doi.org/10.1007/978-981-33-4976-6_7

large capacity. One approach to accommodate services with varying requirements is to use network slicing technologies [8, 19, 20], whereby multiple virtual network slices are constructed over a physical network. By constructing one network slice for each service, the network can satisfy the service requirements.

Traffic conditions can change with time and affect the service quality of the networks. For instance, an increase in the amount of traffic causes congestion, resulting in delays. Overprovisioning is one approach to satisfy the requirements even in the case of traffic changes. With this method, the network operators can satisfy the requirements by allocating a sufficient amount of resources to account for traffic changes. However, this requires a large amount of resources, thereby incurring a substantial cost.

Therefore, the dynamic control of resources is required [1, 2, 17], in which the requirements can be satisfied with a limited amount of resources by dynamically allocating the resources to follow traffic changes. Using this approach, resources are allocated to those areas and services that require more resources and are released if they are no longer required. Resource allocation can be performed based on the observed traffic conditions. However, resource allocation based on observed traffic does not correspond to the actual traffic when significant traffic changes occur. Instead, the configured resource allocation is not changed until the next control cycle. This problem can be addressed by setting a short control interval; however, this approach may cause network destabilization.

Resource allocation based on prediction is a method for accommodating fluctuating traffic without causing network destabilization [13, 14]. In this approach, a controller collects traffic information and predicts future traffic. Thereafter, the controller proactively allocates resources based on the predicted traffic conditions.

Prediction accuracy is important for resource allocation based on prediction, as inaccurate prediction of the traffic amount can result in improper resource allocation and congestion. Numerous methods have been proposed to predict future traffic conditions [7, 11, 12, 15, 18]. In the majority of these methods, future traffic prediction is performed based on the time series of the monitored traffic; that is, future traffic is predicted from previously monitored traffic. However, it is difficult to achieve accurate traffic prediction based only on previously monitored traffic if the signs of fluctuations are not included in the previously monitored traffic.

Real-world information is useful for the prediction of future traffic. This information can include signs of traffic changes that are not included in previously monitored data. For example, knowledge of the number of users related to a service in each area can improve the prediction accuracy of the service traffic, as traffic in an area increases if the number of users in the area increases. An increase in the number of users in adjacent areas is also a sign of an increase in users and traffic related to the service. For IoT services, such real-world information can be obtained from IoT devices. However, it is challenging to model the relationship between real-world information and future traffic. Therefore, a method is required that can predict future traffic conditions using information whose relationship with future traffic cannot be clearly modeled.

In this chapter, I present the application of Yuragi learning, which is inspired by the cognitive processes of the human brain and makes decisions using uncertain information. Numerous studies have attempted to formalize human decision-making. One model for human decision-making is the Bayesian decision-making theory, which treats observed information and the confidence of cognitive objects as stochastic variables. The variables are updated by Bayesian inference every time a new observation is obtained. The brain then makes decisions based on the stochastic variables. Using the above process, the brain can make a decision even if the observed information is unclear. Yuragi learning is a method inspired by the abovementioned cognitive processes of the human brain, and we apply it to predict future traffic conditions based on real-world information and control resource allocation [3, 16]. This chapter describes the application of Yuragi learning to predictive network control based on our work [3, 16].

7.2 Predictive Network Control Based on Yuragi Learning

This section discusses the application of Yuragi learning to predictive network control. I present the model of the cognitive processes of the human brain that inspired Yuragi learning and then discuss our application of the model to predictive network control based on real-world information.

7.2.1 Model of Human Cognition

Yuragi learning is inspired by the model of cognitive processes in the human brain based on Bayesian decision-making theory [5]. Figure 7.1 presents an overview

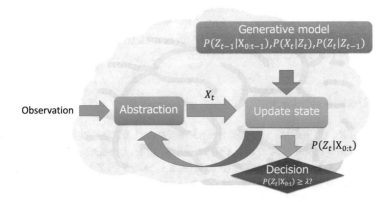

Fig. 7.1 Model of human cognition

of the model. The model encodes the predefined i options $\phi_1, \cdots, \phi_i$ known as an attractor and makes decisions depending on the option of the current status. The decision state Z_t is the internal state of the model, where Z_t is updated by performing Bayesian inference whenever a new observation is obtained.

7.2.1.1 Abstraction

In this model, whenever a new observation is obtained, the observation is abstracted. The abstracted observation is represented by the vector X_t.

7.2.1.2 Generative Model

This model includes the following generative model for the decision state Z_t and observation X_t:

$$Z_t - Z_{t-\Delta_t} = \Delta_t f(Z_{t-\Delta_t}) + \sqrt{\Delta_t} w_t$$

$$X_t = M\sigma(Z_t) + v_t,$$

where $f(Z)$ represents the Hopfield dynamics, w_t and v_t are Gaussian noise variables, $M = [\mu_i, \cdots, \mu_N]$ is a matrix containing the observation values, and μ_i is the observation value corresponding to the state ϕ_i, which is the ith predefined attractor. Furthermore, $\sigma(x)$ is a sigmoid function $\frac{\tanh(ax/2)+1}{2}$, where a is the slope of the function.

7.2.1.3 Update of State

This model updates the decision state Z_t every time X_t is obtained by inverting the generative model using Bayesian inference. Because the generative model is nonlinear, Bitzer et al. used the unscented Kalman filter [9] to update the decision state of Z_t. In addition to updating the decision state, the posterior distribution $P(Z_t|X_t)$ over the decision state is obtained.

7.2.1.4 Decision-Making

The above state estimation outputs the posterior probability $P(Z_t|X_t)$. Thus, the decision is made based on the probability. Bitzer et al. introduced the threshold λ. When $P(Z_t = \phi_i) > \lambda$, the option ϕ_i is selected. When $P(Z_t = \phi_i) \leq \lambda$ for all i, no decision is made until a new observation is obtained.

7.2.2   Application of Yuragi Learning to Predictive Network Control

7.2.2.1   Overview

Traffic conditions have an effect on the service quality of networks; for example, an increase in the amount of traffic causes congestion, which worsens the delay. Resource allocation should thus be varied to follow changes in traffic conditions. One approach for allocating resources to follow traffic condition changes is dynamic resource allocation based on observed information. However, resource allocation based on observed traffic does not correspond to actual traffic when significant traffic changes occur. Instead, the configured resource allocation is not changed until the next control cycle. This problem can be solved by setting a short control interval; however, this may cause network destabilization.

Therefore, we require a method to allocate resources based on traffic prediction [13, 14] and refer to this approach as *predictive network control*. Figure 7.2 presents an overview of this approach. A controller is deployed that periodically obtains observations and predicts future traffic conditions. It calculates a suitable network configuration that satisfies the requirements of the predicted future traffic; thereafter, it configures the network. By repeating these steps, predictive network control proactively allocates resources.

Traditional methods for predicting future traffic are based on the time series of the monitored traffic. However, it is challenging to achieve accurate traffic prediction based only on previously monitored traffic if the fluctuation signs are not included in the previously monitored traffic. Although real-world information is useful for the prediction of future traffic, the relationship between real-world information and future traffic is difficult to model. Therefore, we apply Yuragi learning to predict future traffic conditions.

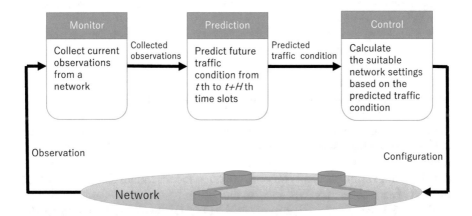

Fig. 7.2 Overview of predictive network control

Figure 7.3 presents an overview of predictive network control based on Yuragi learning. To apply Yuragi learning, we first define the decision-making options using previously monitored observations. In addition to the options, we store the network configuration corresponding to each option considering the future traffic conditions. By storing the corresponding configurations, we can obtain a suitable network configuration by determining the option that the current state belongs to.

Predictive network control based on Yuragi learning operates as follows. The controller periodically obtains observations, including both traffic information and real-world information, related to and collected via the service. Thereafter, it abstracts the observations into X_t. Using X_t, the controller updates its state, Z_t. Finally, the controller determines a suitable network configuration based on the probability of the current state $P(Z_t|X_{t:0})$, where $X_{t:0} = (X_0, X_1, \cdots, X_t)$. The remainder of this subsection explains the details of predictive network control based on Yuragi learning.

7.2.2.2 Options and Network Configuration Corresponding to Each Option

In predictive network control, we first define the decision options and the network configuration corresponding to each option. These are defined offline using the traffic and real-world information monitored previously. We denote the set of period of time t included in the monitored information $T^{monitor}$ and denote the observations at time t by vector O_t. We also denote the traffic demands at time t by vector D_t. O_t includes D_t.

We first define the options using O_t for $t \in T^{monitor}$. We divide the set of O_t into multiple clusters so that each cluster includes similar observations, as illustrated in Fig. 7.4. Thereafter, we define one option for each cluster. That is, the controller determines the cluster to which the current condition belongs whenever a new observation is obtained. Any method can be used to divide the set of O_t into clusters, such as k-means++ [4].

Subsequently, we define the network configuration corresponding to each option. In network control, the network configuration is calculated so that the possible traffic demands can be accommodated without violating the service requirements. Thus, it is necessary to define the possible traffic demands for each option. In predictive network control, the possible traffic demands are set considering the future traffic demands to allocate resources proactively. To define the possible traffic demands, the variation in the traffic demands must be considered; even if the current status defined by the observable information is the same, the future traffic demands may differ. One approach to consider the various future traffic demands is to define the possible future traffic demands for each option as the maximum demands for all of the data in each cluster corresponding to the option.

Figure 7.5 presents the steps to obtain a suitable network configuration for the ith option. We first extract the data included in the cluster corresponding to the ith option, C_i. Thereafter, we calculate the possible future traffic demands for each

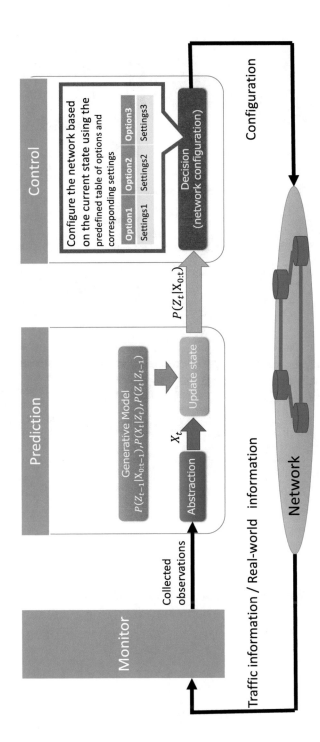

Fig. 7.3 Overview of predictive network control based on Yuragi learning

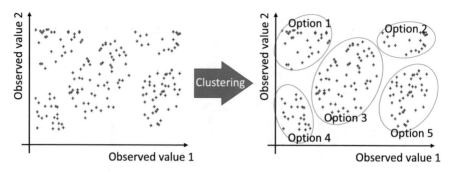

Fig. 7.4 Creating options using previously obtained information

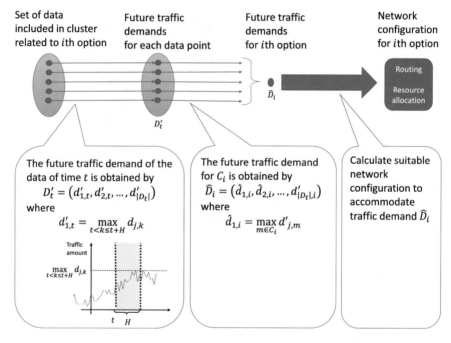

Fig. 7.5 Creating network configuration for each option

of the extracted data points. The possible traffic demand for the data of time t is obtained by

$$D'_t = \left(d'_{1,t}, d'_{2,t}, \cdots, d'_{|D_t|,t} \right),$$

where

$$d'_{j,t} = \max_{t \le k \le t+H} d_{j,k}.$$

In the above equation, $d_{j,k}$ is the jth element of D_k, and H is the distance to the predictive horizon. Using D'_m, we can obtain the possible traffic demands for option i as follows:

$$\hat{D}_i = \left(\hat{d}_{1,i}, \hat{d}_{2,i}, \cdots, \hat{d}_{|D_t|,i} \right),$$

where

$$\hat{d}_{j,i} = \max_{t \in C_i} d'_{j,t},$$

where C_i is the set of period of time in which the observations belong to the cluster corresponding to the ith option. Finally, we set the network configuration for option i so that the traffic demand $\hat{D}_i$ can be accommodated without violating the service requirements. Any method can be used to calculate the network configurations, such as a method for solving optimization problems [13].

7.2.2.3 Abstraction in Predictive Network Control

In predictive network control based on Yuragi learning, whenever the controller obtains new observations O_t, it first calculates vector X_t that abstracts the observation O_t. Any method can be used for abstracting the observations. One approach is to set X_t to O_t. In addition, X_t can be set using a machine learning method. In this chapter, we set X_t using the k-nearest neighbor algorithm. In this method, we store the previously monitored observations and the options corresponding to the observations. When observation O_t is obtained, the controller compares O_t with the stored observations and retrieves k observations that are nearest to O_t. Thereafter, X_t is obtained by

$$X_t = \left(x_{t,1}, x_{t,2}, \cdots, x_{t,S} \right),$$

where S is the number of options and $x_{t,i}$ is the ratio of retrieved observations whose corresponding options are i. X_t obtained by the above steps indicates how close the current observation is to the observations of each option.

7.2.2.4 Generative Model in Predictive Network Control

We use the same generative model as that used by Bitzer et al. [5]. That is, the generative models of the decision state Z_t and observation X_t are

$$Z_t - Z_{t-\Delta_t} = \Delta_t f(Z_{t-\Delta_t}) + \sqrt{\Delta_t} w_t$$

$$X_t = M\sigma(Z_t) + v_t,$$

where $f(z)$ represents the Hopfield dynamics, w_t and v_t are Gaussian noise variables, and $\sigma(x)$ is a sigmoid function. The ith element of X_t represents the ratio of the nearest observations whose options are i. That is, if the current option is i, X_t is a vector whose ith element is 1 and the other elements are 0. Therefore, we set M as an identity matrix.

7.2.2.5 Update in Predictive Network Control

The controller updates the decision state Z_t every time X_t is obtained by inverting the generative model using Bayesian inference in a similar manner to the work by Bitzer et al. [5].

7.2.2.6 Decision-Making in Predictive Network Control

The above procedure outputs the posterior probability $P(Z_t|X_{t:0})$. We perform resource allocation based on this confidence.

If the current observed information clearly indicates that the current state corresponds to a certain option, the value of $P(Z_t|X_{t:0})$ is high only for that option. However, a case may exist in which $P(Z_t|X_{t:0})$ is high for multiple options. Therefore, we set priorities for the options in advance according to the allocated resources; that is, we set a high priority for options that allocate more resources. Thereafter, we select the option with the highest priority among the options with a corresponding $P(Z_t|X_{t:0})$ that is larger than a given threshold to avoid the risk of lacking resources.

7.3 Hierarchical Predictive Network Control Based on Yuragi Learning: Resource Allocation Among Network Slices

7.3.1 Overview

In this section, we discuss predictive network control for IoT services. We assume that one network slice is constructed for each IoT service. Two types of resource allocation are available: resource allocation within each slice and among slices. Resource allocation within a slice is changed by varying the traffic routes within the slice and replacing virtual machines, for example. However, resource allocation among slices is changed by allocating additional resources to the congested network slices and releasing redundant resources from other network slices.

Predictive control based on Yuragi learning described in Sect. 7.2 is applicable to resource allocation within each slice. The controller periodically obtains the traffic and real-world information related to and collected via the IoT service and changes the resource allocation based on Yuragi learning using the collected information. However, it is difficult for a single controller to manage the entire network, as a large amount of real-world information is related to one IoT service. Moreover, the operators of IoT services may not wish to share real-world information related to their service with others.

Therefore, in this section, we consider the case in which multiple controllers cooperate to manage resource allocation within and among network slices. An overview is presented in Fig. 7.6. We deploy two types of controller: a network slice controller, which manages resources within the corresponding network slice, and a resource allocation controller, which performs resource allocation among the slices.

Each network slice controller manages resources based on Yuragi learning, as discussed in Sect. 7.2. That is, the network slice controller periodically collects real-world information related to the IoT service and updates its state using the

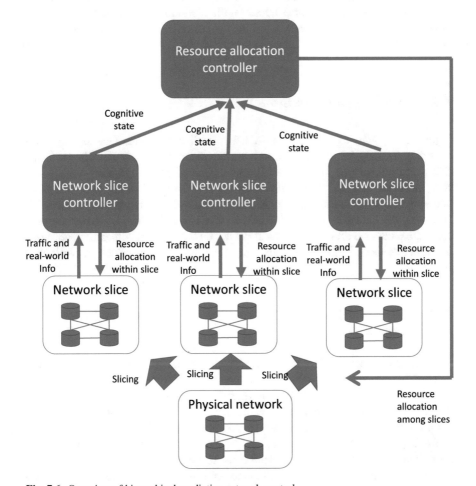

Fig. 7.6 Overview of hierarchical predictive network control

collected information. The network slice controller identifies the current condition according to its state. However, resource allocation within a network slice cannot be determined only by the identified condition. The network slice controller must also consider the resources allocated to its corresponding network slice. Thus, the network slice controller determines the network configuration within the network slice based on the identified current condition and allocated resources.

The resource allocation controller manages resources among the network slices and collects the cognitive states of the network slice controllers instead of real-world information. In this manner, the resource allocation controller identifies the current condition of all slices without collecting real-world information. Using the collected information, the resource allocation controller updates its state, similarly to the network slice controller, and identifies the current condition. The resource allocation controller stores the options related to the conditions and the resource allocation corresponding to each condition. Thereafter, the resource allocation controller allocates the resources based on the identified condition. The remainder of this section describes the behavior of the network slice controller and the resource allocation controller.

7.3.2 Network Slice Controller

Figure 7.7 presents an overview of the network slice controller. One network slice controller is deployed for each network slice, and it periodically collects traffic and real-world information and updates its state. Thereafter, it configures the network slice based on its state. The behavior of the network slice controller is based on predictive network control based on Yuragi learning, as mentioned in Sect. 7.2. However, the network slice controller controls the network slice based not only on its state but also on the resources allocated by the resource allocation controller. In the remainder of this subsection, I describe the behavior of the network slice controller.

7.3.2.1 Options and Network Configuration Corresponding to Each Option

The options are defined by the previously monitored observations in a similar manner to predictive network control based on Yuragi learning described in Sect. 7.2. Thereafter, the network configuration corresponding to each option is defined. Unlike the case explained in Sect. 7.2, the network configuration is not defined only for the current condition; a suitable configuration of the network slice is also dependent on the allocated resources. Therefore, the network configurations are defined for a pair of options and allocated resources.

As explained in Sect. 7.2, we define the options by dividing the set of observations O_t into multiple clusters so that each cluster contains similar observations.

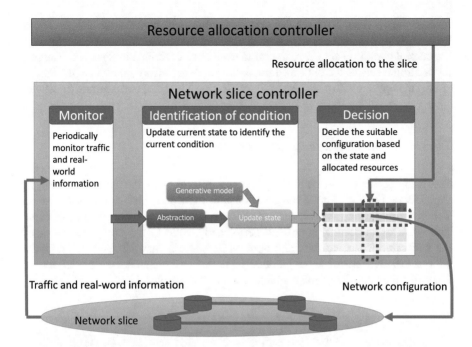

Fig. 7.7 Behavior of slice controller

Subsequently, we define the possible traffic demands for option i, $\hat{D}_i$, in the same manner as described in Sect. 7.2. In addition to the possible traffic demands, we obtain the set of resource allocations to the slice. Then, we set the network configuration for option i and allocated resources R so that the traffic demand $\hat{D}_i$ can be accommodated without violating the service requirements using the allocated resources R.

7.3.2.2 Identification of Current Condition

The network slice controller identifies the current conditions based on Yuragi learning and uses the same steps to identify the current condition as those described in Sect. 7.2. That is, the network slice controller periodically obtains observations and abstracts them into X_t. By using X_t, the controller updates its state, Z_t. Finally, the network slice controller obtains the probability of the current state, $P(Z_t|X_{t:0})$.

7.3.2.3 Configuration of Network Slice

The network slice controller determines the network configurations based on $P(Z_t|X_{t:0})$ and the currently allocated resources R_t. Because the currently allocated resources are provided, the controller can determine suitable network configurations by identifying the options related to the current condition. The options related to the current condition are set in the same manner as described in Sect. 7.2. Namely, the controller selects the option whose corresponding traffic demands are the largest among the options with a corresponding $P(Z_t|X_{t:0})$ that is above a certain threshold.

7.3.3 Resource Allocation Controller

The resource allocation controller allocates resources to the network slices, as depicted in Fig. 7.8. The resource allocation controller also operates based on Yuragi learning using the collected cognitive states of the network slice controllers. The resource allocation controller periodically collects the cognitive states of the network slice controllers and updates its state. It then changes the resource allocation to the network slices based on its state. In the remainder of this subsection, I describe the behavior of the resource allocation controller.

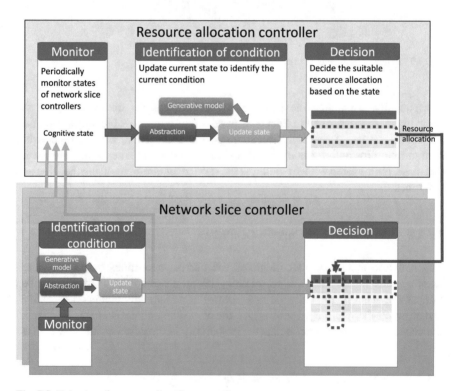

Fig. 7.8 Behavior of resource allocation controller

7.3.3.1 Observations in Resource Allocation Controller

The resource allocation controller collects the cognitive states of the network slices. Information regarding the cognitive state of the network slice n at time t is represented by

$$o_{t,n} = \left(\frac{P(Z_{t,n} = \phi_{1,n})}{\max_j P(Z_{t,n} = \phi_{j,n})}, \frac{P(Z_{t,n} = \phi_{2,n})}{\max_j P(Z_{t,n} = \phi_{j,n})}, \ldots, \frac{P(Z_{t,n} = \phi_{S_i,n})}{\max_j P(Z_{t,n} = \phi_{j,n})} \right),$$

where $Z_{t,n}$ is the state of network slice n at time t, $\phi_{j,n}$ is the state corresponding to the jth option in network slice n, and S_i represents the number of options in the network slice controller for network slice i. Furthermore, $o_{t,n}$ is a vector in which the element corresponding to the options with the largest confidence is 1.

The resource allocation controller collects $o_{t,n}$ from all network slice controllers. Thus, the observation of the resource allocation controller O_t is expressed as

$$O_t = \left(o_{t,1}, o_{t,2}, \ldots, o_{t,N} \right).$$

7.3.3.2 Options and Network Configuration Corresponding to Each Option

To define the options and network configurations corresponding to each option, we obtain the time series of the information regarding the observations and the traffic demands in advance. We can obtain this information by operating the network slice controller in an actual environment or by performing a simulation. We denote the traffic demand of the network slice n at time t as $D_t^{(n)}$.

After obtaining the information, we define the options by dividing the observations into clusters in the same manner as described in Sect. 7.8. Thereafter, we define the resources allocated to the slices for each option. Even if the current configuration of the network slice requires additional resources, the network slice controller can accommodate all demands without violating the service requirements by changing the network slice configurations. Therefore, we define the resource allocation for each option using the overall information of the network.

We first define the possible traffic demands for each option for all network slices in the same manner as in Sect. 7.8. That is, the possible traffic demands of network slice n for the ith option of the resource allocation controller are obtained as follows:

$$\hat{D}_i^{(n)} = \left(\hat{d}_{1,i}^{(n)}, \hat{d}_{2,i}^{(n)}, \cdots, \hat{d}_{|D_i|,i}^{(n)} \right),$$

where

$$\hat{d}_{j,i}^{(n)} = \max_{t \in C_i} \max_{t \geq k \geq t+H} d_{j,k}^{(n)}.$$

In the above equation, C_i is the set of periods of time in which the observations belong to the cluster corresponding to the ith option, $d_{j,k}^{(n)}$ is the jth element of $D_k^{(n)}$, and H is the distance to the predictive horizon. In this way, we obtain the required resources for each network slice to accommodate all traffic demands $\hat{D}_i^{(n)}$. Any method can be used to obtain the required resources. One approach is to solve an optimization problem to set the configurations of the network slices to accommodate all traffic demands of all network slices. We can also use heuristic methods to obtain the resource allocations.

7.3.3.3 Identification of Current Condition

Similarly to the network slice controller, the resource allocation controller identifies the current conditions based on Yuragi learning. Thus, the resource allocation controller periodically collects the cognitive state of all network slice controllers and abstracts the observations into X_t. Using X_t, the controller updates its state, Z_t. Finally, the network slice controller obtains the probability of the current state $P(Z_t|X_{t:0})$.

7.3.3.4 Resource Allocation

The resource allocation controller allocates resources to network slices based on the current state $P(Z_t|X_{t:0})$. We allocate resources defined for the option with the largest confidence $P(Z_t = \phi_i)$.

7.4 Simple Example

In this section, I provide a simple example of predictive network control based on Yuragi learning.

7.4.1 Scenario

7.4.1.1 Network

In this example, we ran the resource allocation controller and network slice controller in a simple network, as illustrated in Fig. 7.9. This network was constructed using core nodes and a gateway. The core nodes were connected to base stations, and in this network, each device was connected to a base station. The traffic from each device was sent to the Internet via the base station, the core node connected to the base station, and the network constructed from the core nodes. Traffic to a device was sent via the opposite route.

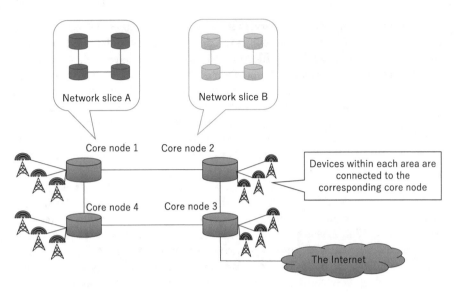

Fig. 7.9 Simple network

In this example, we divided the capacity of the links into the network slices. That is, the network slices were constructed over the physical network of the core nodes. The total capacity of the link between node 3 and the gateway was set to a sufficiently large value, and the total capacity of the links between nodes 2 and 3 and nodes 3 and 4 was set to 25 Gbps. The capacity of the other links was set to 10 Gbps.

7.4.1.2   Network Slices

In this example, we accommodated the following two network slices:

Slice A: This was the network slice for the connected vehicles. This service required low latency. For the sake of simplicity, we set the service requirements of this slice to accommodate traffic along the shortest path.

Slice B: This was the network slice for a video streaming service for mobile users. The latency did not have a significant effect on this service; thus, this service required high capacity.

7.4.1.3   Traffic and Real-World Information

In this example, we generated traffic and real-world information using the pseudo-generated Global Positioning System (GPS) trajectory dataset known as Open PFLOW [10]. This dataset includes the position of each person every 5 min in the

Tokyo metropolitan area. In addition to the position, this dataset includes the mode of transportation of each person, such as walking, bicycle, car, or train. Based on this mode, we set the number and position of devices connected to each slice. However, the number of people included in this dataset is 617,040, which does not include data corresponding to all people in the metropolitan area. Therefore, assuming that multiple users move similarly, we generated the number of devices in each area by assigning a scale factor to each user in Open PFLOW and summing the scaled number of users in the area. We generated multiple datasets by randomly changing the scale factors. We used one dataset as training data for prediction and the other dataset for evaluation.

In this example, we focused on an area in Shinjuku-ku, Tokyo, whose size was 4 × 4 km. We divided the area into four subareas, and the traffic from each subarea was sent via the same core node. We generated traffic for each slice as follows. The traffic of Slice A was generated based on the number of vehicles within each subarea, and the traffic of Slice B was generated based on the number of people within each subarea. For both slices, we generated the amount of traffic from a device based on the estimated Internet traffic in Japan according to the Ministry of Internal Affairs and Communications [6]. The estimated Internet traffic included information regarding the amount of mobile traffic to/from each user at each time in 1 day. We generated the amount of traffic from each device at each time using this information. Figure 7.10 presents the amount of traffic generated for this example. The figure indicates that the traffic changes in Slice A differed from those in Slice B; the traffic peak in Slice A was at approximately noon, whereas the traffic peak in Slice B was in the evening. Therefore, this simple example demonstrates that our method successfully allocated resources following traffic changes.

7.4.1.4   Controller Settings

Network Slice Controller

In this example, each network controller used the traffic amount to/from each area and the number of devices in each area that were related to each slice as the observations. We defined the options using the observations. In this example, the network slice controller for Slice A had seven options, whereas that for Slice B had three options.

We then defined the possible traffic demands for each option by setting H to 1 h and used the quantized traffic demands with a step size of 5000 Mbps. Tables 7.1 and 7.2 display the quantized possible traffic demands for each option of Slices A and B.

In this example, the network slice controller determined traffic routes between the core nodes and the gateway. Thus, we set the routes for each pair of options and allocated resources. In this example, we set the routes based on the constrained shortest path first (CSPF) algorithm. With this algorithm, we first sorted the traffic in descending order in terms of volume. Then, we calculated the traffic routes of high-volume traffic. For each traffic instance, we calculated the shortest paths after removing links that lacked sufficient capacity to accommodate the traffic route to be

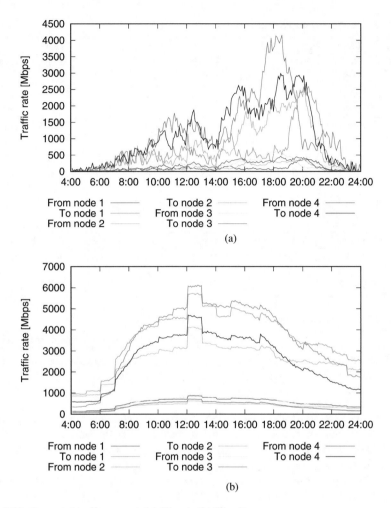

Fig. 7.10 Generated traffic amount. (**a**) Slice A. (**b**) Slice B

calculated. In this manner, we were able to identify routes that could accommodate all traffic.

The network slice controller identified the current state according to Yuragi learning. The generative model in Yuragi learning used in this case included two parameters: the standard deviations of v and w. In this example, we set these parameters to the values indicated in Table 7.3.

Resource Allocation Controller

The resource allocation controller allocated resources to network Slices A and B. The observations of the resource allocation controller were the cognitive states of

Table 7.1 Traffic amount predicted for each option (Slice A) [Mbps]

Option	From Node 1	To Node 1	From Node 2	To Node 2	From Node 3	To Node 3	From Node 4	To Node 4
0	5000	5000	5000	5000	5000	5000	5000	5000
1	5000	5000	5000	5000	5000	10, 000	5000	10, 000
2	5000	5000	5000	5000	5000	10, 000	5000	5000
3	5000	5000	5000	5000	5000	5000	5000	10, 000
4	5000	10,000	10, 000	5000	5000	10, 000	5000	10, 000
5	5000	10,000	10, 000	5000	5000	5000	5000	10, 000
6	5000	10,000	5000	5000	5000	5000	5000	10, 000

Table 7.2 Traffic amount predicted for each option (Slice B) [Mbps]

Option	From Node 1	To Node 1	From Node 2	To Node 2	From Node 3	To Node 3	From Node 4	To Node 4
0	5000	5000	5000	5000	5000	5000	5000	5000
1	5000	10, 000	5000	5000	5000	10, 000	5000	5000
2	5000	5000	5000	5000	5000	10, 000	5000	5000

Table 7.3 Parameters of generative model used in example

	Network slice controller	Resource allocation controller
Standard deviation of v	0.30	0.10
Standard deviation of w	0.20	0.10

the network slice controllers. We defined the options using the observations. In this example, the resource allocation controller contained five options.

We defined the resources allocated to the network slices in advance for each option. In this case, we defined the resources allocated to the network slices using the possible traffic demands for both network slices. Thus, we calculated the traffic routes for both Slices A and B by the CSPF algorithm. We assigned priority to Slice A because this slice required the traffic to be accommodated using the shortest path. Therefore, we calculated the traffic route in Slice A first and then calculated the traffic route in Slice B. Thereafter, we obtained the resources required for allocation to each slice by calculating the required resources to accommodate the traffic in each slice through setting the routes calculated by the above steps. Table 7.4 displays the resource allocation for each option.

The resource allocation controller identified the current state based on Yuragi learning. The generative model in Yuragi learning used in this chapter included two parameters: the standard deviations of v and w. In this example, we set these parameters to the values indicated in Table 7.3.

Table 7.4 Resource allocation in each option [Mbps]

Option	Slice	1–2	1–4	2–1	2–3	3–2	3–4	4–1	4–3
0	A	5000	0	5000	10, 000	10, 000	5000	0	5000
	B	5000	0	5000	10, 000	10, 000	5000	0	5000
1	A	5000	0	5000	10, 000	10, 000	5000	0	5000
	B	5000	0	5000	10, 000	10, 000	5000	0	5000
2	A	5000	0	5000	10, 000	5000	5000	0	5000
	B	5000	5000	5000	10, 000	10, 000	5000	0	15, 000
3	A	5000	0	5000	10, 000	10, 000	5000	0	5000
	B	5000	5000	5000	10, 000	10, 000	5000	0	15, 000
4	A	5000	0	5000	5000	5000	5000	0	5000
	B	5000	0	5000	10, 000	10, 000	5000	0	15, 000
5	A	5000	0	5000	5000	5000	5000	0	5000
	B	5000	0	5000	15, 000	15, 000	15, 000	5000	15, 000

7.4.2 Results

Figure 7.11 presents the time series of confidence $P(Z_t = \phi_i)$. In this figure, the horizontal axis indicates the time, while the vertical axis indicates the confidence of each option. The figure represents the changes in the confidence of each option following the changes in the condition. In the network slice controller for Slice A, the confidence of the option corresponding to the case in which the traffic amount was small was high during the early time slots. Thereafter, the confidence of the options corresponding to the case in which the traffic amount was large increased following the increase in traffic. Finally, the confidence of the options corresponding to the case in which the traffic amount was small increased again following the decrease in traffic. In the network slice controller for Slice B, the confidence of the option also changed following the traffic changes.

Figure 7.12 presents the time series of the confidence $P(Z_t = \phi_i)$ in the resource allocation controller. In this figure, the horizontal axis indicates the time, while the vertical axis indicates the confidence of each option. The figure indicates that the confidence of each option also changed in the resource allocation controller following the changes in the condition. The resource allocation controller identified that both network slices required a small number of resources during the early time slots. Thereafter, the confidence of the options corresponding to the case in which Slice B required additional resources increased. In the evening, the number of vehicles increased. Following the increase in traffic from vehicles, the confidence of the options corresponding to the case in which Slice A required additional resources increased.

Figure 7.13 presents the resources allocated at each time and indicates the capacities of each link allocated to Slices A and B. The resource allocation in the links between nodes 1 and 2 did not change because the link capacities were limited. However, the resource allocation in the other links changed. The resources allocated

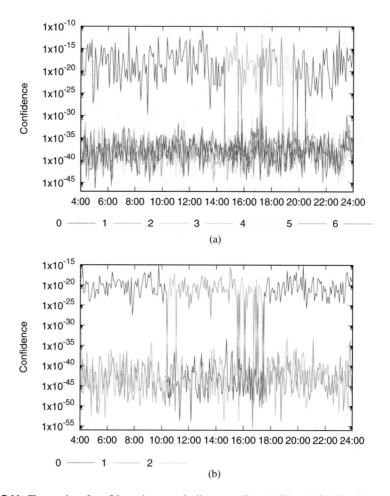

Fig. 7.11 Time series of confidence in network slice controller. (**a**) Slice A. (**b**) Slice B

to Slice A increased at approximately noon to accommodate the peak traffic in Slice A. The resources of the links from nodes 4 to 1, nodes 2 to 3, nodes 3 to 2, and nodes 3 to 4 allocated to Slice A increased. However, the resources of the link from nodes 3 to 2 allocated to Slice B increased in the evening following the increase in traffic in Slice B. Moreover, we verified that all traffic demands could be accommodated by the routes set by the network slice controllers. That is, the controllers were able to allocate sufficient resources before the traffic changes.

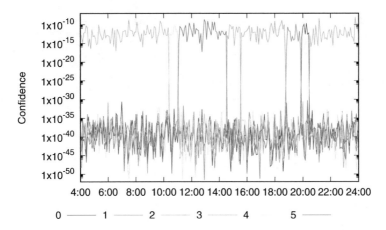

Fig. 7.12 Time series of confidence in resource allocation controller

7.5   Conclusion

In this chapter, we discussed the application of the Yuragi learning mechanism to the control of networks accommodating IoT services. As IoT devices have become increasingly popular, numerous services have been provided through networks. The traffic resulting from these services must be accommodated to satisfy the service requirements. One approach to accommodate the traffic is the use of network slicing, which provides multiple network slices for the network services. Resources for each slice should be dynamically allocated following traffic changes. Predictive network control is a method for accommodating fluctuating traffic without degrading the service quality. Predictive network control requires predicted future traffic, for which real-world information is useful. In this chapter, we considered the use of real-world information related to IoT services to predict future traffic. However, it is difficult to model the relationship between real-world information and future traffic. Therefore, we applied Yuragi learning inspired by the cognitive processes of the human brain that make decisions based on uncertain information. Thus, in this chapter, I presented the application of Yuragi learning to predictive network control based on real-world information.

Acknowledgments The research introduced in this chapter was supported in part by contract research for the National Institute of Information and Communications Technology (NICT), "Research and development on new network infrastructure technology creating the future."

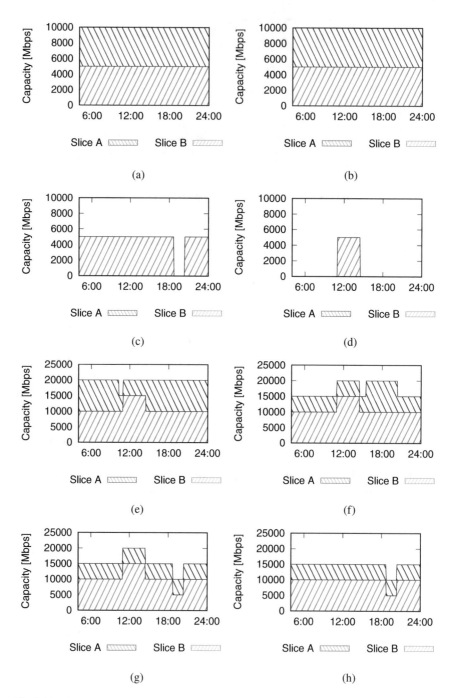

Fig. 7.13 Time series of resource allocation of each link. (**a**) Link 1–2. (**b**) Link 2–1. (**c**) Link 1–4. (**d**) Link 4–1. (**e**) Link 2–3. (**f**) Link 3–2. (**g**) Link 3–4. (**h**) Link 4–3

References

1. Agarwal, S., Kodialam, M., Lakshman, T.: Traffic engineering in software defined networks. In: INFOCOM 2013, pp. 2211–2219. IEEE, Piscataway (2013)
2. Akyildiz, I.F., Lee, A., Wang, P., Luo, M., Chou, W.: A roadmap for traffic engineering in SDN-OpenFlow networks. Comput. Net. **71**, 1–30 (2014)
3. An, S., Ohsita, Y., Murata, M.: Resource allocation control considering quality of service to multiple network slices using human brain cognition model. In: Forum on Information Technology (in Japanese). IPSJ/IEICE, Tokyo (2019)
4. Arthur, D., Vassilvitskii, S.: k-means++: The advantages of careful seeding. Technical report, Stanford (2006)
5. Bitzer, S., Bruineberg, J., Kiebel, S.J.: A Bayesian Attractor Model for Perceptual Decision Making. PLoS Comput. Biol. **11**(8), e1004442 (2015)
6. Computer Communications Division, Telecommunications Business Department, Telecommunications Bureau, Ministry of Internal Affairs and Communications: Aggregation and provisional calculation of internet traffic in Japan (2018)
7. Feng, H., Shu, Y.: Study on network traffic prediction techniques. In: International Conference on Wireless Communications, Networking and Mobile Computing, vol. 2, pp. 1041–1044. IEEE, Piscataway (2005)
8. Foukas, X., Patounas, G., Elmokashfi, A., Marina, M.K.: Network slicing in 5G: Survey and challenges. IEEE Commun. Mag. **55**(5), 94–100 (2017)
9. Julier, S.J., Uhlmann, J.K.: New extension of the Kalman filter to nonlinear systems. In: Signal Processing, Sensor Fusion, and Target Recognition VI, vol. 3068, pp. 182–194. International Society for Optics and Photonics, Bellingham (1997)
10. Kashiyama, T., Pang, Y., Sekimoto, Y.: Open PFLOW: Creation and evaluation of an open dataset for typical people mass movement in urban areas. Transport. Res. C Emerg. Technol. **85**, 249–267 (2017)
11. Lu, W.: Parameters of network traffic prediction model jointly optimized by Genetic Algorithm. JNW **9**(3), 695–702 (2014)
12. Otoshi, T., Ohsita, Y., Murata, M., Takahashi, Y., Ishibashi, K., Shiomoto, K.: Traffic prediction for dynamic traffic engineering. Comput. Netw. **85**, 36–50 (2015)
13. Otoshi, T., Ohsita, Y., Murata, M., Takahashi, Y., Kamiyama, N., Ishibashi, K., Shiomoto, K., Hashimoto, T.: Traffic engineering based on model predictive control. IEICE Trans. Commun. **98**(6), 996–1007 (2015)
14. Otoshi, T., Ohsita, Y., Murata, M., Takahashi, Y., Ishibashi, K., Shiomoto, K., Hashimoto, T.: Hierarchical model predictive traffic engineering. IEEE/ACM Trans. Netw. **26**(4), 1754–1767 (2018)
15. Rutka, G.: Neural network models for Internet traffic prediction. Elektronika ir Elektrotechnika **68**(4), 55–58 (2015)
16. Satake, K., Ohsita, Y., Murata, M.: Predictive traffic engineering incorporating real-world information inspired by the cognitive process of the human brain. In: International Conference on Information and Communication Technology Convergence (ICTC), pp. 543–548. IEEE, Piscataway (2019)
17. Tayyaba, S.K., Akhunzada, A., Amin, N.U., Shah, M.A., Khan, F., Ali, I.: NPRA: novel policy framework for resource allocation in 5G software defined networks. EAI Endorsed Trans. Mob. Commun. Appl. **4**(13) (2018)
18. Yu, Y., Song, M., Fu, Y., Song, J.: Traffic prediction in 3G mobile networks based on multifractal exploration. Tsinghua Sci. Technol. **18**(4), 398–405 (2013)
19. Zhang, H., Liu, N., Chu, X., Long, K., Aghvami, A.H., Leung, V.C.: Network slicing based 5G and future mobile networks: mobility, resource management, and challenges. IEEE Commun. Mag. **55**(8), 138–145 (2017)
20. Zhou, X., Li, R., Chen, T., Zhang, H.: Network slicing as a service: enabling enterprises' own software-defined cellular networks. IEEE Commun. Mag. **54**(7), 146–153 (2016)

Chapter 8
Another Prediction Method and Application to Low-Power Wide-Area Networks

Daichi Kominami

Abstract The Internet of Things has become increasingly widespread, and low-power wide-area (LPWA) technology has attracted attention as one of its elemental technologies. LPWA technology achieves wide-area communication without consuming a large amount of energy, which facilitates various types of applications for sensing and collecting data. LoRa (Long Range) is a type of LPWA communication technology that uses unlicensed bands. Because it is possible to build a self-managed network with LoRa, many services using LoRa are scattered in the same area without an administrator. As a result, the communication performance of LoRa may be degraded due to unintended radio interference. However, because many LPWA techniques, including LoRa, have a low data rate, it is difficult to gather sufficient control information to avoid the degradation of communication performance. In this chapter, we propose a method for predicting the network congestion state based on Yuragi learning that enables prediction from fluctuating and noisy data by successive Bayesian estimation. Through computer simulation, we demonstrate that the network state can be predicted by our proposed method with a small amount of control information.

8.1 Introduction

Low-power wide-area (LPWA) networks, which realize low-power and wide-area communication, have attracted increasing attentions [10], as LPWA techniques facilitate the development of the Internet of things (IoT). Using LPWA techniques, it is possible to easily collect data from users and devices. SigFoX, a representative LPWA standard, has been available in more than 65 countries since 2019. Another representative standard, LoRa (Long Range), is available in more than 100 countries, and the LoRa Alliance was launched in 2015 and has more than 500 member

D. Kominami (✉)
Graduate School of Information Science and Technology, Osaka University, Suita, Osaka, Japan
e-mail: d-kominami@ist.osaka-u.ac.jp

© Springer Nature Singapore Pte Ltd. 2021   181
M. Murata, K. Leibnitz (eds.), *Fluctuation-Induced Network Control and Learning*,
https://doi.org/10.1007/978-981-33-4976-6_8

companies. One company in each country is allowed to deploy the SigFox network service as the network operator, and users must use the public network operated by the operator. With LoRa, in contrast, users can freely build their own networks using products standardized by the LoRa Alliance.

LPWA technology achieves low-power consumption and allows IoT devices to operate for many years on a single battery charge. In addition, signals can be delivered to the gateway from a distance of approximately 10 km in the absence of obstacles. By combining a communication module that supports LPWA technology with various sensing devices, it is possible to easily collect information from devices. Although LPWA technology is a combination of the existing technologies, it has high applicability and can facilitate the development of IoT [8].

In a wide-area network (WAN) using LPWA devices, a star network consisting of a gateway and nodes is generally constructed. In other words, a many-to-one communication system between nodes and a gateway is adopted. In particular, uplink communication from a node to the gateway is widely used in IoT and constitutes most of the traffic. Because nodes are not required to relay data, they only need to start the wireless module at the time at which they wish to transmit data, and can switch off the wireless module at other times. Such intermittent communication significantly reduces the power consumption of the wireless module. LPWA technology is also designed to have a high link budget, which is one of the factors enabling long-distance communication. The elemental technologies of an LPWA network include the use of a relatively low frequency band, a modulation method that is effective against interference and noise, and an antenna with high reception sensitivity. Particularly, the data rate is designed to be relatively low, that is, several hundred bps to several kbps, which is lower than that of a conventional mobile network.

In LPWA communication, the data transmission time is long because the data rate is low. In LoRa and SigFox, the ALOHA protocol is used for the medium access control layer, and there is concern that data frame collisions increase as the number of nodes increases. By verifying the received signal strength before transmitting a signal, it may be possible to avoid collisions by detecting a carrier radio wave, as used in IEEE 802.11. However, if a clear channel assessment (CCA) threshold of approximately −80 dB as in IEEE 802.11 is used, collisions are expected to occur as in the case of the ALOHA protocol, as radio signals with a signal strength lower than the threshold can reach the gateway due to the high antenna reception sensitivity of LPWA nodes. However, the lower the threshold is, the greater the probability is that nodes will judge that the wireless channel is busy, which will reduce the number of transmission opportunities.

Using LoRa, users can build their own private networks with nodes and gateways [5]. Therefore, in the future, there will be environments in which many private LoRa wide-area networks (LoRaWAN) are constructed close to each other. However, the influence of interference increases due to an increased number of nodes [12]. In addition to interference, the communication quality between nodes and the gateway fluctuates over different time scales due to various causes, such as the existence of a system using the same frequency band and the occurrence of obstacles. In LoRa, multiple data rates and wireless channels are available, and the gateway can handle fluctuations in communication quality by assigning an appropriate data rate and wireless channel to nodes. However, to determine an appropriate data rate or wireless channel according to the change in communication quality, it is necessary for the gateway to detect the change in communication quality and make a decision accordingly.

In this chapter, we propose a method for LoRaWAN in which the gateway autonomously detects changes in communication quality and assigns a wireless channel to be used for each node. Here, the wireless communication quality fluctuates temporally and spatially; however, nodes do not always have a Global Positioning System (GPS) module and are not always able to observe the communication quality of available wireless channels. Furthermore, the wireless resources available for information collection are limited, and a long time would be necessary to gather sufficient information to perform optimal channel assignment. There have been various attempts to overcome channel fluctuations in LPWA networks [1]; however, most of the approaches involve the physical layer.

Therefore, we focus on the information recognition mechanism of the human brain, which performs appropriate inferences even when only limited information is available. In the human brain, top-down information processing occurs that makes decisions by comparing input information from various sensory organs with memories in the brain. It has recently been reported that this series of information processing can be explained by a decision-making model based on Bayesian inference [4]. In the Bayesian attractor model (BAM) proposed in [9], a hidden variable (decision variable) representing the decision state of the brain is defined on the state space and follows the dynamics with multiple attractors. In addition, a nonlinear function for converting the state space of the decision variable into the feature space is defined. Feature variables are defined in the feature space, and each feature variable corresponding to each attractor corresponds to a memory in the brain, as described above. The BAM models information processing in the brain by (1) observing sensory information, (2) updating the decision variables by Bayesian inference based on the observed information and the dynamics of the decision variables, and (3) making a decision based on the posterior probability distributions of the decision variables.

In this chapter, we assume that multiple LoRaWANs exist in the same area, and the decision variables in the BAM are associated with the degree of communication congestion in the wireless channel. An overview of our proposal is illustrated in Fig. 8.1. Each attractor represents a different degree of congestion and is mapped to feature values at each degree of congestion by conversion using the nonlinear function described above. Our proposal based on the BAM is operated on a gateway (or network server ahead of it); thus, the feature values must be observed by the gateway. The gateway periodically calculates the feature values based on communication with nodes and inputs the calculated feature values to the BAM. Then, the decision variables are updated according to the input. By performing this for each wireless channel, it is possible to estimate the degree of congestion of each wireless channel. Here, because the decision variables are updated using

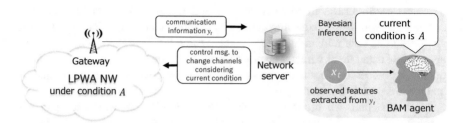

Fig. 8.1 Proposal overview: autonomous control loop with the Bayesian attractor model (BAM)

Bayesian estimation, they are obtained as posterior probability distributions. A value indicating how close a decision variable is to each attractor (i.e., the *confidence*) is defined based on the posterior probability distribution. When the confidence levels of all the wireless channels exceed the threshold values, wireless channel assignment control is performed. We define the congestion degree as a state in which N LPWA nodes exist. Thus, the number of LPWA nodes can be determined as the degree of congestion of each wireless channel. In wireless channel assignment control, the number of nodes is equalized for each wireless channel.

Our contributions in this chapter are summarized as follows. We first demonstrate that the prediction method based on the BAM, called Yuragi learning, can effectively predict the wireless channel congestion. To predict the communication congestion in LoRaWAN, the data reception rate and data decoding success rate, which are realistically available to the gateway, are used as features. Second, the Bayesian filter used in the BAM is changed from an unscented Kalman filter (UKF) to a particle filter to facilitate the design of state estimation by the BAM. In the original BAM, the prediction accuracy may be reduced when the number of memories is greater than the dimension of features; however, it can be improved using a particle filter.

The remainder of this chapter is organized as follows. In Sect. 8.2, we present the details of the BAM and our extension of its filter. In Sect. 8.3, we present the definition of feature values and describe how we determine the memories embedded in attractors of the BAM. In addition, we describe the wireless channel assignment method. In Sect. 8.4, we evaluate the performance of our proposal through the computer simulation. In Sect. 8.5, we provide the conclusions.

8.2 Bayesian Attractor Model

The BAM is a model of information perception, discrimination, and decision-making in the brain using a Bayesian estimation framework. Our research group has demonstrated that the BAM can be used to reconfigure a virtual network over an optical network to create a virtual network suitable for a given traffic situation [6]. In addition, we have demonstrated that by using the BAM for rate control in MPEG

Dynamic Adaptive Streaming over HTTP (MPEG-DASH), it is possible to achieve rate selection according to available bandwidth and user preference [3].

The BAM consists of three main components: (i) information perception, (ii) information discrimination, and (iii) decision-making. In (i) information perception, the BAM expresses observed information with a feature vector. In (ii) information discrimination, the BAM compares the perceived information with past experience and memory and determines whether it matches any of them. Past experience and memory are provided as a vector with the same dimension as the input feature vector. Here, the BAM defines the internal state of the brain as following nonlinear dynamics with K attractors, and past experiences and memories are associated with each attractor. Based on this dynamics and perception of information, the internal state of the brain is updated based on Bayesian inference. Because the internal state of the brain follows dynamics that have attractors, the internal state approaches one of the attractors. In (iii) decision-making, the confidence is derived based on the probability density distribution of the internal state of the brain obtained in information discrimination. If the confidence value exceeds a predefined threshold, it is deemed that the perceived information matches past experience or memory corresponding to the internal state.

In the BAM, variables that represent the internal decision state of the brain are defined as hidden variables that are updated according to known dynamics. The decision variables z approach one of the attractors existing in the state space due to the aforementioned dynamics. The BAM estimates z based on the perceived information; however, because z is a hidden variable, Bayes' theorem is used to estimate z.

8.2.1 Generative Model

In the BAM, the decision variables z of the brain are represented by a random variable, and z is updated by nonlinear dynamics with K attractors. Given an initial state, z approaches one of the attractors according to Eq. (8.1).

$$z_t = z_{t-\Delta} + \Delta g(z_{t-\Delta}) + \sqrt{\Delta} \boldsymbol{w}_t. \tag{8.1}$$

Here, z is a vector of dimension K, Δ is an update interval, and $\boldsymbol{w}_t$ is a random number following a normal distribution $\mathcal{N}(0, \ q^2/\Delta)$, where q represents the magnitude of the process error included in the generative model called the dynamics uncertainty. g represents winner-take-all network dynamics and is defined as follows:

$$g(z) = k \left(L\sigma(z) + b^{lin}(\phi - z) \right). \tag{8.2}$$

Here, k is a constant that determines the scale of the update, and ϕ is a matrix of $K \times 1$ whose elements all have the same value ϕ_g. b^{lin} indicates the strength of a goal state attractor. In addition, $L = b^{lat}(I - \mathbf{1})$, where I is a unit matrix, $\mathbf{1}$ is a matrix whose elements are all 1 (in each case, the size is $K \times K$), and b^{lat} indicates the strength of the lateral inhibition in this winner-take-all network dynamics. σ is a sigmoid function, $\sigma(z_i) = 1/(1 + e^{-d(z_i - o)})$, and each element z_i of z is converted into the range [0 1]. d represents the attenuation characteristic, and o represents the position of the inflection point of the sigmoid function. By repeating the dynamics g, only one element of z converges to ϕ_g. By setting $o = \phi_g/2$ and $b^{lat}/b^{lin} = 2\phi_g$, the other elements of z converge to $-\phi_g$. In other words, K attractors in the dynamics of z are of dimension K, where only the ith element is ϕ_g, while the other elements are $-\phi_g$ ($i = 0, \ldots, K - 1$).

In the BAM, each attractor of the generative model is associated with a feature vector representing past memory and experience. A feature vector x_t corresponding to a certain decision variable is generated as follows.

$$x_t = M\sigma(z_t) + v_t. \tag{8.3}$$

Here, $M = [\mu_0 \ \mu_1 \ \ldots \ \mu_{K-1}]$ is a feature matrix containing the feature vectors where μ is an m-dimensional vector. M is an $m \times K$ matrix, and v_t is a random number following the normal distribution $N(0, r^2)$, where r represents the sensory uncertainty.

8.2.2   State Estimation by Bayesian Filters

The BAM estimates the decision variables z based on the predefined generative model and perceived information. Because z is a hidden variable and is updated temporally according to the generative model, a sequential Bayesian filter is used for estimation. Reference [9] uses a UKF [13], which is a type of Bayesian filter. A general Kalman filter has poor estimation performance when handling nonlinear dynamics. In contrast, the UKF is a method that alleviates the shortcomings of the Kalman filter and approximates the probability distribution using a generative model and a small number of samples called sigma points calculated based on the estimated standard deviation. By using the UKF, the posterior probability distribution $P(z_t|x_t)$ of z at time t is obtained, and the probability density of each attractor $P(z_t = \phi_n|x_t)$ can be calculated. The authors of [9] refer to this probability density value as the *confidence*. Confidence is used instead of marginal likelihood as a decision-making index in the BAM because the amount of computation required for marginalization increases exponentially when the number of attractors increases. When the confidence is greater than the predefined threshold λ, ϕ_n is determined as the estimation result. Here, ϕ_n is a matrix of $K \times 1$, and in this chapter, the nth element and other elements are ϕ_g and $-\phi_g$, respectively ($n = 0, \ldots, K - 1$). Note that ϕ_n must be carefully defined so that each element of ϕ_n corresponds to each attractor. Because the UKF is used in the BAM, the confidence is defined with the density function of the multivariate normal distribution. Therefore, the confidence of the BAM exponentially decreases as the number of dimensions of z increases. Therefore, an appropriate value of λ must be carefully considered in advance.

From an engineering perspective, the BAM can be regarded as an estimation or prediction tool for determining the coincidence between features observed from noise sources and features memorized in advance. However, there are several challenges to be resolved when applying the BAM in this context. In general, when the dimension of z is larger than the dimension of x, the estimation of z by x is an underdetermined problem, that is, the solution is not uniquely determined. In the BAM, z is updated by nonlinear dynamics with attractors, and therefore, z is expected to eventually converge to one of the attractors. However, the Kalman filter minimizes the variance of the model error; thus, once z is estimated at a position other than an attractor and the estimated variance at that time has a small value, z will stop at an equilibrium point other than an attractor. At equilibrium points other than attractors, the confidence value is low, and it is uncertain whether the confidence exceeds the threshold.

Bitzer et al. [9] evaluated the characteristics of the BAM as an information processing model of the brain. In this study, they did not consider situations in which the dimension of z was equal to or larger than that of x, or where the estimation of z became an underdetermination problem. In this study, we use a particle filter [7], which is a type of Bayesian filter, instead of the UKF to enable estimation even in the sub-determined case. The particle filter is a state estimation method based on Bayesian estimation. The probability distribution required for state estimation is represented as a set of many particles rather than a mathematical expression, and the posterior probability distribution is approximated by weighted particles in the state space. The particle filter performs a sequential Monte Carlo simulation and provides weights to particles based on the likelihood of observations. By defining a likelihood function that has a lower value at non-attractor positions, we can expect z to have a higher probability of approaching attractors than remaining at a non-attractor equilibrium point described above, and the estimated result will be more likely to be in the vicinity of one of the attractors.

The algorithm of the particle filter applied to the BAM is as follows. Here, it is assumed that all particles p_i $(i = 0 \ldots N_P - 1)$ are initialized in advance.

1. Update p_i according to Eq. (8.1): $p_i \leftarrow p_i + \Delta g(p_i)$.
2. Calculate the weight of a particle i: $W_i = P(y|p^i)$, where y denotes the observed data. If the distribution of the observed value y is known, it is used for the likelihood function $P(y|p^i)$. If it is unknown, an approximation with an appropriate distribution is used.
3. Calculate $\hat{z}$ that is given as a weighted average of p_i: $\hat{z} = \sum_{i=0}^{N_P-1} \mathcal{W}^{-1} W_i p_i + w_t$, where $\mathcal{W} = \sum_{i=0}^{N_P-1} W_i$.
4. Resample p_i using a sequential importance resampling (SIR) method [11]. In the SIR method, the current particle set is replaced by a new particle set that

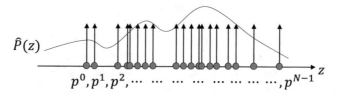

Fig. 8.2 Estimation of probability density with a particle filter

consists of N_P particles selected from the current particle set with probabilities proportional to their weight W_i.

The estimation accuracy generally improves by increasing the number of particles N_P; however, the calculation time increases proportionally to N_P (Fig. 8.2).

8.2.3 Comparison of Bayesian Filters in the Bayesian Attractor Model

We compared the properties of the BAM using a particle filter (BAM-PF) and the BAM using the UKF (BAM-UKF). To this end, we generated N_s random numbers, $S = s_0, s_1, s_2, \ldots, s_{N_s-1}$, that followed a normal distribution $\mathcal{N}(\mu, d^2)$ and input the numbers to the BAM. To investigate the changes in confidence of both types of BAM, we changed the mean value of the normal distribution.

We set $N_s = 600$, and $M = (1, 2, 3)$. Here, s_0–s_{199} followed a normal distribution with $\mu = 1$ and $d = 1$, s_{200}–s_{399} followed a normal distribution with $\mu = 2$ and $d = 1$, and s_{400}–s_{599} followed a normal distribution with $\mu = 3$ and $d = 1$. Then, we set the number of attractors to 3 and stored M in the attractors. An input s_t was given at each time step t.

BAM-UKF

Table 8.1 presents a list of parameters for the BAM-UKF and their values. In [9], k was set to 4; however, when we set k to 4, the decision variables approached one of the attractors very slowly. Therefore, we set k to 40 so that the decision variables rapidly approached to the attractors. It should be noted that when we set k to a value much larger than 40, the decision variables did not leave an attractor once approached it.

Here, the decision variable z had three elements: $z = [z_1\ z_2\ z_3]^T$. For both the BAM-UKF and BAM-PF, the number of attractors was 3 (ϕ_1, ϕ_2, ϕ_3), and they were defined as $[10, -10, -10]^T$, $[-10, 10, -10]^T$, $[-10, -10, 10]^T$, respectively.

BAM-PF

We calculated the confidence of the BAM-PF as for the BAM-UKF. First, we set a larger value of k for the BAM-PF than for the BAM-UKF. This is because the particle filter was based on randomly generated particles, which increased the

Table 8.1 Parameters of the Bayesian attractor model using the unscented Kalman filter (BAM-UKF)

Parameter	Value	Description	Same value used in [9]
q	0.1	Dynamics uncertainty	✓
r	$d = 1$	Sensory uncertainty	
b^{lat}	1.7	Strength of lateral inhibition	✓
b^{lin}	$b^{lat}/20$	Strength of goal state attractor	✓
Δ	0.004	Input interval	✓
k	40	Scale of dynamics	
g	10	Distance factor between attractors	✓
r	0.7	Slope of sigmoid function	✓
o	$g/2$	Center of sigmoid function	✓

variability of z itself compared to that of the BAM-UKF. Therefore, we set k to 1,000. In the BAM-PF, it was also necessary to determine the number of particles and the likelihood function. In this evaluation, the number of particles was set to 1,000, and the likelihood function L of particle p_i was defined by Eq. (8.4), where y denotes the observation.

$$L(y|p_i) = \mathcal{N}(y - M\boldsymbol{\sigma}(p_i), d). \qquad (8.4)$$

Results

Figures 8.3 and 8.4 illustrate the transition of the decision variables and confidence levels of the BAM-UKF, respectively, while Figs. 8.5 and 8.6 illustrate those of the BAM-PF.

In the BAM-UKF, the decision variables remained at non-attractor points. In contrast, in the BAM-PF, decision variables remained closer to the attractors than in the BAM-UKF. Because the confidence increased as the decision variables approached an attractor, the confidence of the attractor that memorized the value closest to the observed value was higher in the BAM-PF. In the BAM-UKF, the decision variables did not always approach the attractors, which made it difficult to set an appropriate certainty threshold for decision-making. However, in the BAM-PF, this situation did not occur because the decision variables approached an attractor.

8.3 Methods for Channel Congestion Prediction and Channel Assignment

In this section, we present methods for predicting changes in the degree of congestion of wireless channels and for assigning wireless channels to nodes based on the predictions.

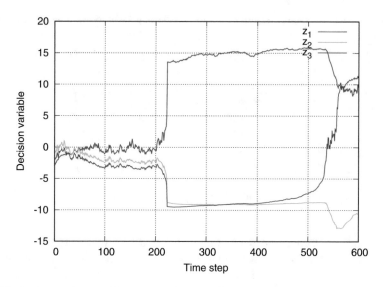

Fig. 8.3 Decision variable z of Bayesian attractor model using the unscented Kalman filter (BAM-UKF)

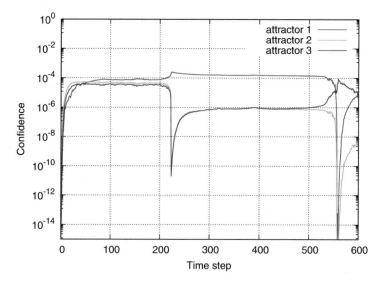

Fig. 8.4 Confidence $P(z = \phi_n|y)$ of Bayesian attractor model using the unscented Kalman filter (BAM-UKF)

We consider a communication system using LoRa, and aim to achieve efficient communication even when the quality of wireless communication changes unexpectedly. There are various situations in which the communication quality changes, such as an increased number of private networks using LoRa, an increased number

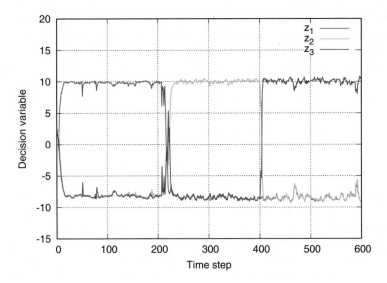

Fig. 8.5   Decision variable z of Bayesian attractor model using the particle filter (BAM-PF)

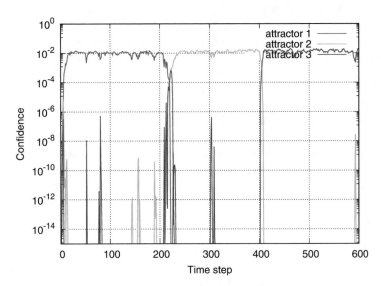

Fig. 8.6   Confidence $P(z = \phi_n | y)$ of Bayesian attractor model using the particle filter (BAM-PF)

of devices that use the same wireless frequency band, and the existence of structures that disrupt communication. To identify the type of situation, information with sufficient granularity and quantity is required. However, as mentioned in Sect. 8.1, LPWA communication systems have limited wireless resources available to collect information for control.

If a feature value is unique when an event S occurs, we can estimate whether the event is S by storing this feature value in the BAM. Although it is not necessarily the case that the event is S for a feature, in this section, we assume that the relationship between features and events is one-to-one. In addition, as a matter of course, it is difficult to acquire feature values for an event before the event occurs. It is also possible that unforeseen events may occur during network operation. In this situation, it is necessary to store new features in the BAM during network operation. The problem, however, is that the network system can observe the features but not the event itself. Therefore, the network administrator should either observe or set the event manually. We assume that unique features are obtained for individual events and that the correspondence between events and features is known. However, the update of the number of attractors and features is beyond the scope of this chapter.

Because users can build private LoRaWANs, multiple networks can coexist in the same area, and the degree of congestion of the wireless channel can change dynamically. In this case, we predict the change in the degree of congestion for each wireless channel using the BAM. The BAM stores features μ_i representing the degree of congestion when there are N_i nodes (where i is the number of states to remember, which matches the number of attractors). In this section, we use the data reception rate and data decoding success rate, which are realistically available to the gateway, as feature variables. At regular intervals, these values within the last interval are calculated and input to the BAM. The data reception rate is the value obtained by dividing the number of data that are actually received by the gateway and successfully decoded during the interval, by the expected number of received data. Here, it is assumed that the gateway has a known data transmission schedule for each node. The data decoding success rate is the percentage of data received during the interval that are successfully decoded.

Wireless channels are assigned to nodes so that the degree of congestion of each wireless channel is identical. This is because the probability that randomly generated data transmission does not collide with each other exponentially decreases with an increase in the degree of congestion. Here, because the gateway knows the number of nodes belonging to the same network, it estimates N_i, the number of nodes belonging to the same network and other networks, using the BAM. When the number of nodes on the wireless channel c obtained from the result of the BAM is $N_i(c)$, the number of nodes using the wireless channel c on the same network of the gateway is $N_{same}(c)$, the number of available wireless channels is N_c, and the number of nodes to be assigned to each wireless channel, $N'_{same}(c)$, is determined as follows:

$$N'_{same}(c) = \frac{1}{N_c} \sum_{c} N_i(c) - \{N_i(c) - N_{same}(c)\}. \tag{8.5}$$

The wireless channel assignment of a node is performed after the confidence levels of all wireless channels observed by the BAM exceed the threshold value. After the confidence levels of all channels exceed the threshold value, the gateway assigns the channel of each node using a piggyback mechanism. When the gateway transmits an acknowledgment (ACK) message for data from a node, it contains a control instruction for the node to change its assigned channel specified by the gateway.

The wireless channels are reassigned to all nodes. In the proposed method, we select nodes that have not been reassigned in a random order and assign c to them until the number of allocated wireless channels for channel c is $N'_{same}(c)$. Because the nodes are not assigned a new channel until they transmit data after the confidence levels exceed the threshold, the next assignment is not performed for a certain period of time after the assignment.

In addition, no features can be obtained for a wireless channel that does not have any nodes. Although unused wireless channels can be examined by methods other than the BAM, in this section, we define the minimum value N_{min} for $N'_{same}(c)$. If $N'_{own}(c)$ is less than N_{min}, N'_{own} is set to N_{min}. In this way, we assign wireless channels so that the degree of congestion of channels other than c is identical.

8.4 Evaluation

8.4.1 Simulation Settings

In this section, we present a simulation for evaluating our proposal in a LoRaWAN scenario. 200 LoRa nodes and one gateway were installed in an area of $5 \times 5 \text{ km}^2$. The x and y coordinates of the nodes were determined according to uniform random numbers in the range of 0–5 km, and the gateway was set at (0, 0). Each node periodically generated data once in every 300 s interval. The data generation timing was asynchronous among nodes, and a uniform random number from -2.5 to 2.5 s was added to the interval to avoid continuous data collision. The number of wireless channels that were available for nodes was set to 4, and at the start of the simulation, the number of nodes that used each wireless channel was set to 50.

The data and ACK frame sizes were 50 bytes and 10 bytes, respectively, and the data rate was 1.5 kbps. The nodes and gateway performed carrier sense for 5 ms before transmitting the data and an ACK frame. The CCA threshold was set to -83 dB, and frame transmission was cancelled when a signal using the same channel was detected. In data transmission, a node retransmitted data only once in total when the wireless channel was detected to be busy by carrier sense or an ACK was not returned from the gateway.

For signal propagation, we considered only direct waves and modeled them by the Friis transmission equation [2]. Here, the used frequency was 920 MHz, the attenuation coefficient was 2.5, the antenna reception sensitivity of the nodes and the gateway was -131 dB, the transmission power was 13 dB, and the gain of the transmitting and receiving antenna was 5 dB. For data and ACK, decoding errors were stochastically generated according to the signal-to-noise ratio (SNR) at the time of reception. In this simulation, we used a simple decoding error model because the bit error rate differed depending on the environment. Specifically, the bit error rate was 100% when the SNR was 0 dB or less, 50% when the SNR was 0–5 dB, 10% when the SNR was 5–10 dB, 1% when the SNR was 10–20 dB, and 0% when SNR was 20 dB or more.

The total simulation time was 36,000 s. Every 12,000 s, 50 additional LoRa nodes were added at random positions only on a specific wireless channel. These added nodes did not belong to the network that the above-mentioned nodes and gateway belonged to. The added nodes generated data at the same interval (300 s); however, the gateway did not return an ACK to the additional nodes.

The parameters for the BAM-PF were the same as in Sect. 8.2.3. As the degree of congestion, we derived features in advance by simulation where the number of nodes using a certain wireless channel was 50, 100, and 150. The features were stored in the attractors. The feature matrix M used in the simulation is as follows:

$$M = \begin{pmatrix} 0.98\ 0.97 \\ 0.95\ 0.94 \\ 0.91\ 0.90 \end{pmatrix},$$

where the first column is the data reception rate and the second column is the data decoding success rate. The first, second, and third rows are the values when the number of nodes was 50, 100, and 150, respectively.

For the likelihood function L of particle p_i, a multivariate normal distribution was used as an approximate distribution of the observed values. Then, $L(y|p_i) = \mathcal{N}(y - M\sigma(p_i), \Sigma)$ and Σ is a variance–covariance matrix of observations. For this, we used a variance–covariance matrix of features when the number of nodes using a certain wireless channel was 100, which was obtained by simulation in advance. The variance–covariance matrix Σ used in the simulation is as follows:

$$\Sigma = \begin{pmatrix} 0.0043\ 0.0033 \\ 0.0033\ 0.0037 \end{pmatrix}.$$

The gateway calculated features (i.e., the data reception rate and data decoding success rate) every minute and input them to the BAM-PF immediately after the calculation. The minimum number of nodes N_{min} assigned to each wireless channel was set to 20.

8.4.2   Simulation Results

We first present the results when the gateway did not perform channel assignment. Figures 8.7, 8.8, and 8.9 present the sequence of the observation input, decision variable, and confidence of the wireless channel to which nodes were added. The results indicate that the BAM-PF was able to predict changes in the degree of congestion. Here, the threshold of the confidence was set to 10^{-3} based on Fig. 8.9.

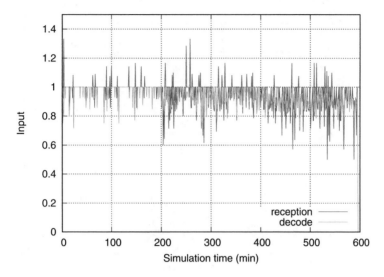

Fig. 8.7 Observation input

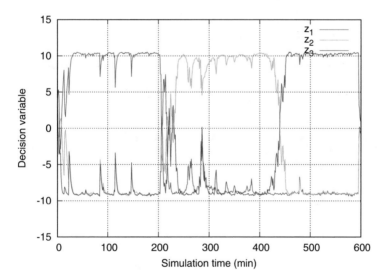

Fig. 8.8 Decision variable z

Then, the average time step from the time of node addition to the time at which the confidence levels of the correct attractor exceeded the threshold and had the maximum value among the confidences of three attractors was 31.0 over 100 simulation trials. When the threshold of the confidence was set to $5 \cdot 10^{-4}$, the average time step became 18.3.

Compared with the results presented in Sect. 8.2.3, the changes in the decision variables and confidence are more unstable. This is because although we used

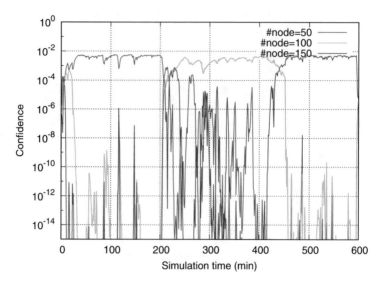

Fig. 8.9 Confidence $P(z = \phi_n | y)$

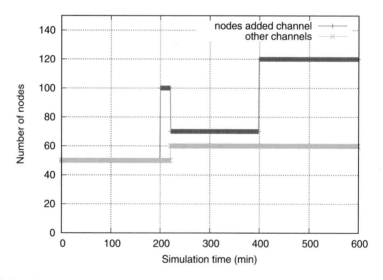

Fig. 8.10 Wireless channel assignment

a likelihood function that was based on a multivariate normal distribution, both distributions of the data reception rate and decoding success rate did not follow a normal distribution, as indicated in Fig. 8.7.

Finally, we present the number of nodes in each wireless channel when the gateway performed the channel assignment. The simulation results are presented in Fig. 8.10. We set the threshold to 10^{-3}.

All wireless channels had a degree of congestion with 50 nodes until 13,200 s (220 min). When the simulation time reached 13,200 s (220 min), 50 nodes were added to the wireless channel c, where $N'_{same}(c)$ defined in Eq.(8.5) became 12.5. Because N_{min} was 20, the number of nodes in channel c became 20. Then, 60 nodes were assigned to each wireless channel that had no additional nodes. When the simulation time reached 24,000 s (400 min), another 50 nodes were added to the wireless channel. Thereafter, because calculating $N'_{same}(c)$ resulted in -25, the number of nodes in channel c remained 20. The results in Fig. 8.10 demonstrate that our intended control was realized. However, we also identified cases in which the expected results were not achieved when the confidence did not exceed the threshold. In the future, the operating parameters of the BAM will be examined in greater detail.

8.5 Conclusion

In this chapter, we present a method for estimating the degree of wireless channel congestion in LoRaWAN using the BAM, which is a human cognition model. For the Bayesian filter used in the BAM, we use a particle filter instead of the UKF to more easily apply the BAM to LoRaWAN scenarios. By using the BAM, our proposal can predict changes in the degree of wireless channel congestion with a comparatively small amount of information without being greatly affected by temporary fluctuations in the observed values. In our proposal, the LoRaWAN gateway assigns wireless channels to nodes based on the results of the BAM, which enables adaptive and stable control of the LoRaWAN system. In the future, we will extend the evaluation to clarify parameters that can make the BAM more effective. In addition, we will compare the BAM with other methods to reveal the advantages of the proposed method, especially in terms of communication performance.

References

1. Berg, V., Dore, J.B., Mannoni, V.: Channel estimation strategy for LPWA transmission at low SNR: application to Turbo-FSK. In: Proceedings of IEEE 89th Vehicular Technology Conference (VTC2019-Spring), pp. 1–5. IEEE, Piscataway (2019)
2. Friis, H.T.: A note on a simple transmission formula. Proc. Inst. Radio Eng. **34**(5), 254–256 (1946)
3. Iwamoto, M., Otoshi, T., Kominami, D., Murata, M.: Rate adaptation with Bayesian attractor model for MPEG-DASH. In: Proceedings of IEEE Annual Computing and Communication Workshop and Conference (CCWC), pp. 0859–0865 (2019)
4. Kveraga, K., Ghuman, A.S., Bar, M.: Top-down predictions in the cognitive brain. Brain Cogn. **65**(2), 145–168 (2007)
5. LoRa Alliance: LoRa Specification V1.1 (2017)
6. Ohba, T., Arakawa, S., Murata, M.: A Bayesian-based virtual network reconfiguration in elastic optical path networks. IEICE technical report, PN2016-33 **116**(307), 45–50 (2016)

7. Pitt, M.K., Shephard, N.: Filtering via simulation: auxiliary particle filters. J. Am. Stat. Assoc. **94**(446), 590–599 (1999)
8. Raza, U., Kulkarni, P., Sooriyabandara, M.: Low power wide area networks: an overview. IEEE Commun. Surv. Tutorials **19**(2), 855–873 (2017)
9. Bitzer, S., Bruineberg, J., Kiebel, S.J.: A Bayesian attractor model for perceptual decision making. PLoS Comput. Biol. **11**(8), e1004442 (2015)
10. Sinha, R.S., Wei, Y., Hwang, S.H.: A survey on LPWA technology: Lora and NB-IoT. ICT Express **3**(1), 14–21 (2017)
11. Smith, A.F., Gelfand, A.E.: Bayesian statistics without tears: a sampling–resampling perspective. Am. Stat. **46**(2), 84–88 (1992)
12. Staniec, K., Kowal, M.: LoRa performance under variable interference and heavy-multipath conditions. Wirel. Commun. Mob. Comput. **2018**, 6931083 (2018). https://doi.org/10.1155/2018/6931083
13. Wan, E.A., Van Der Merwe, R.: The unscented Kalman filter for nonlinear estimation. In: Proceedings of the IEEE 2000 Adaptive Systems for Signal Processing, Communications, and Control Symposium, pp. 153–158. IEEE, Piscataway (2000)

Chapter 9
Artificial Intelligence Platform for Yuragi Learning

Toshiyuki Kanoh

Abstract Neural network-based artificial intelligence, such as deep learning, has made remarkable progress by utilizing high-performance computing resources. However, research in neuroscience has revealed that the human brain can realize multimodal, flexible, and efficient cognitive operations by consuming only approximately 21 W of power. This chapter summarizes our challenges in developing an artificial cognitive system based on the Bayesian attractor model and provides a brief introduction to our software prototype called the Yuragi Learning General-Purpose Data Analysis Platform (YGAP).

9.1 Introduction

The human brain weighs only approximately 1.5 kg and consumes approximately 21 W of power. However, it has multimodal, flexible, and efficient cognitive abilities that surpass current supercomputers in everyday tasks. With technological improvements in the temporal and structural resolution of devices that record the activity of the human brain, neuroscientists have made significant progress in understanding how the brain functions. However, it is currently impossible to measure the entire brain at the single neuron level due to the poor spatiotemporal resolution of existing recording devices. In addition, it is impossible to replicate the behavior of the entire brain on a computational device.

Artificial neural networks, such as deep learning networks, use mathematical models to represent the operation of computational neurons and have recently experienced significant progress through the use of high-performance computing resources. In the field of image processing, in particular, deep neural networks achieve similar results to humans and even surpass humans in accuracy [4]. Although artificial neural networks can mimic certain aspects of the brain's

T. Kanoh (✉)
Industry-Academia Collaboration, Graduate School of Information Science and Technology, NEC Brain Inspired Computing Alliance Laboratories, Osaka University, Suita, Osaka, Japan
e-mail: t-kanoh@ist.osaka-u.ac.jp

© Springer Nature Singapore Pte Ltd. 2021
M. Murata, K. Leibnitz (eds.), *Fluctuation-Induced Network Control and Learning*,
https://doi.org/10.1007/978-981-33-4976-6_9

behavior, they are often limited to specific tasks, such as image recognition. Our research goal differs from deep learning by aiming to realize a new artificial intelligence (AI) system that closely follows findings from neuroscience [5]. Recent research has revealed that the human brain can effectively use its limited resources to achieve flexible and high-performance computing tasks. For example, for visual processing, this is achieved by using two distinct cognitive pathways, such as for facial recognition [3] and motion recognition [2]. In addition, in [8], the authors indicated that in the process of acquiring a representation of motion, the representation may be reconstructed using discretized data rather than the original continuous input. To acquire new functions by enhancing the body's capabilities, it has also been reported that the brain can reallocate already utilized resources to extension processing [7]. These findings indicate that the brain follows a strategy that makes the best use of its limited resources. We use knowledge of human cognitive functions to establish computationally executable models with which we replicate human cognitive abilities to develop a brain-inspired cognitive computing system.

The remainder of this chapter is structured as follows. In Sect. 9.2, we summarize the general challenges of realizing a new model of a brain-inspired AI system based on the previous work in [6]. Then, in Sect. 9.3, we describe its implementation, called the Yuragi Learning General-Purpose Data Analysis Platform (YGAP), which utilizes noise for learning and provides an example of its operation in Sect. 9.4.

9.2 Overview of a Brain-Inspired Cognitive Computing System

We first define a brain-inspired cognitive computing system as a system that meets the following requirements:

- optimization of resource usage,
- multimodality for dealing with multiple objectives,
- autonomy to behave in a goal-oriented manner, and
- compliant with explainable AI.

An important piece of information to help optimize utilization of limited resources is the fact that the human brain uses two cognitive pathways. One is a fast and coarse pathway, whereas the other is a slow and fine-grained pathway. For example, the former is used when avoiding danger or defending against attacks while conserving resources under low cognitive load. In contrast, the latter is used for recognizing facial expressions in greater detail when sufficient resources are available. We believe that the efficient use of resources is a major reason why the human brain can effectively operate with much lower power than that of current computers. This mechanism may be helpful in establishing a system that is able to respond to critical situations in real time.

The brain handles different types of sensory input in a similar manner, as the neuron firing process does not depend on sensory perception. The common handling of multimodal sensory input makes it possible to extend incomplete information from one modality by using information from other modalities. Multiple objectives on the basis of multimodal information may also be beneficial for building a robust system in the presence of ambiguous input.

Humans analyze environmental stimuli while decomposing the input into primitive elements. Then, representations of the stimuli are constructed through the combination of the primitive elements depending on the respective goals to understand the environment. It has been reported that even a new representation can be constructed with this combination. Adaptability to a new environment can be used to realize an autonomous adaptive system.

Although current AI technologies, such as deep learning, are gaining popularity, the process is a black box. The features for internal representation are automatically optimized using large amounts of training data, and the visibility of the internal process eventually degrades because human-friendly features are not always extracted. However, for practical use, an externally controllable and transparent system is preferable.

9.2.1 Conceptual Design of a Brain-Inspired Cognitive Computing System

A conceptual diagram of a brain-inspired computing system is presented in Fig. 9.1. It consists of three parts: an encoding module that inputs environmental stimuli as internal information, an integration module that integrates features extracted with

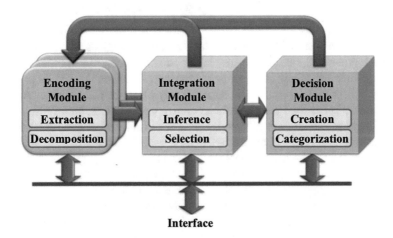

Fig. 9.1 Conceptual design of a brain-inspired cognitive computing system

respect to each modality, and a decision module that handles the manipulation of categories. Each module is composed of submodules.

- The *encoding module* translates system inputs into feature vectors following a common format. First, the module decomposes its inputs into primitive elements. Then, it extracts features from the elements. The format of the vector is independent of the modality.
- The *integration module* integrates the selected element representations. The module infers the element representations from feature vectors while extending noisy or missing information. It selects appropriate representations across multiple modalities according to the system purpose.
- The *decision module* outputs a category that is suitable for the system inputs. This module creates novel categories when necessary, such as when the input is noisy and cannot be categorized.

9.2.2 Architecture Design

We propose a system architecture on the basis of the conceptual diagram in Fig. 9.1, placing emphasis on easily implementing our research results into the system. Research on brain function is progressing rapidly, and it is important for researchers to be able to test their models whenever necessary on a feasible platform. Our proposed platform is presented in Fig. 9.2, and the features of our proposed architecture are as follows:

- event-driven module execution optimized across the entire system,
- data handling structure sharing data via a brain information database (BIDB), and
- graphical user interface (GUI) with internal information monitors and query-based commands for brain information data.

Human brains feature modularity, whereby separate regions work cooperatively with each other, and multiple sensory inputs can be processed in parallel modules. Despite the independent and parallel behavior of the modules without central control, the modules appear to be appropriately optimized as a whole. Applying this optimization method can allow the system to perform more than the summation of the functional modules even though each module is independently designed.

A functional module is driven by specific events and shares data with other modules via the BIDB. A functional module is a process that implements the aforementioned modules illustrated in Fig. 9.1. Event messages, such as the completion of a specific module, are communicated between modules over a messaging bus, and a chain of events forms the processing flow. An event message can be received by multiple modules, and the modules do not share data directly. The data stream is optimized as the entire system communicates between functional modules via the BIDB. Any data can be stored in the BIDB with a history and can be loaded from

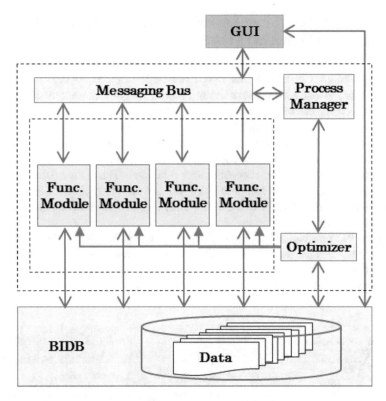

Fig. 9.2 Platform architecture of a brain-inspired cognitive computing system

any module. Indirect communication between modules facilitates the independence of the modules and enables the addition and recombination of modules.

Data communicated between modules can be visualized by a GUI. Any data, such as extracted features, inferred element representations, and category classification history, are stored in the BIDB according to the specifications. Data pertaining to any moment can be accessed; therefore, queries can be issued to observe the internal state and analyze the processes leading to cognition.

The platform makes it easy to implement models of various states and connect each model. We believe that the ease of implementation can enhance brain function research and model implementation. In Sect. 9.3, we present a specific implementation of our proposed concept that utilizes Yuragi learning as the data analysis function.

9.3 Overview of Yuragi Learning General-Purpose Data Analysis Platform (YGAP)

As discussed in Chap. 5 of this book, the brain-inspired cognitive computing model, called Yuragi learning, can categorize various types of multimedia data into predefined categories solely based on their features and without an explicit learning process. Due to the characteristics of the underlying Bayesian attractor model [1], Yuragi learning demonstrates robustness to noise and dispersion of the observed input data. In addition, it has developmental functions, such as automatic knowledge (category) acquisition, and reconstruction functions. Thus, Yuragi learning can autonomously increase and maintain its knowledge. As one application using Yuragi learning, this section introduces the Yuragi Learning General-Purpose Data Analysis Platform (YGAP).

Figure 9.3 presents a functional block diagram of the YGAP. Its architecture is based on a relational database system, and Yuragi learning is implemented as a special data analysis function for classification and categorization in the relational database management system (RDBMS) of the YGAP.

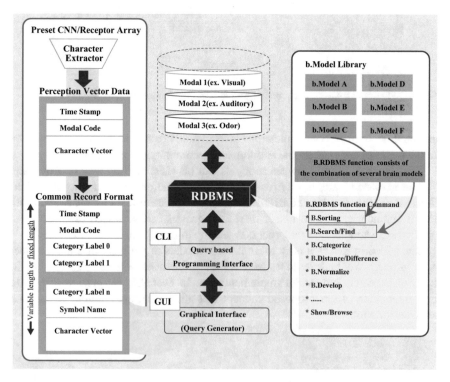

Fig. 9.3 Functional block diagram of Yuragi Learning General-Purpose Data Analysis Platform (YGAP)

The YGAP consists of three components displayed in the left, center, and right parts of Fig. 9.3. The left part represents the application-dependent component consisting of the feature extraction function and data format transformation function. The center part represents the standard RDBMS; in this case, PostgreSQL is used as the base system. The right part represents the special SQL function library using Yuragi learning loaded in the RDBMS.

Figure 9.4 illustrates the interaction between the parts of the system. The left part of the figure represents the client system consisting of the feature extraction function and a web browser, such as the Internet Explorer, while the right part of the figure represents the host system consisting of the YGAP interface module and PostgreSQL with the Yuragi learning function. The users can access the client system through a web browser. The YGAP interface module on the host system provides a GUI that generates SQL queries as PostgreSQL commands and then visualizes the cognition and categorization process as well as the results. The GUI also makes it possible to set the parameters for defining the Yuragi learning characteristics, namely, the dynamics uncertainty, sensory uncertainty, and noise level.

Figure 9.5 presents an overview of the YGAP graphical interface. The GUI provides easily accessible menus for entering parameter values and viewing the graphical output of the results. The YGAP and its installation and quick-start manual can be downloaded from the following URL: https://github.com/nbic-ist-osaka-u-ac-jp/YLPF-Core.

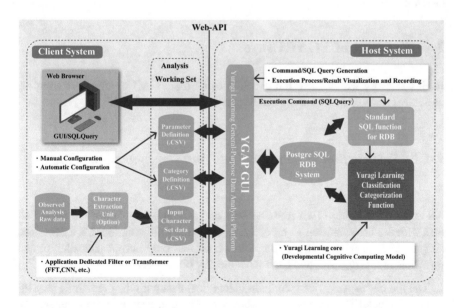

Fig. 9.4 Yuragi Learning General-Purpose Data Analysis Platform (YGAP) system overview

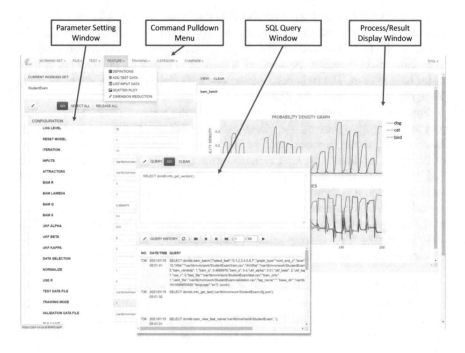

Fig. 9.5 Overview of the Yuragi Learning General-Purpose Data Analysis Platform (YGAP) graphical interface

9.4 Example of YGAP Usage

In this section, we evaluate the capabilities of the YGAP by demonstrating the individual steps of a multi-class classification task with the YGAP.

9.4.1 Preparation of Analysis Data and Configuration of Data Files

All data and definition files must be provided as comma-separated values (CSV) files with the file extension *.csv. Specifically, the following four CSV files must be provided. In its current version, the YGAP can only accept data files in text format as input (Fig. 9.6) and cannot handle raw binary data, such as images.

1. The file category.csv contains categorical data with the predefined classification labels. This example involves a classification task with three categorical types of *cat*, *dog*, and *bird* with values 1, 2, and 3, respectively.

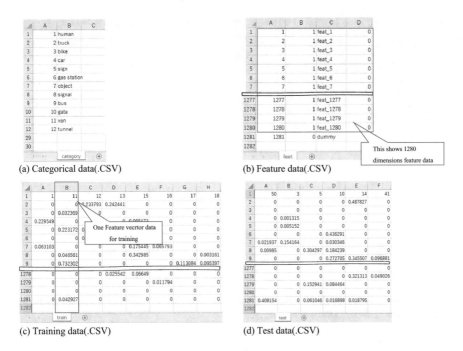

Fig. 9.6 Examples of input files in comma-separated values (CSV) format for Yuragi Learning General-Purpose Data Analysis Platform (YGAP). (**a**) Categorical data (`category.csv`). (**b**) Feature data (`feat.csv`). (**c**) Training data (`train.csv`). (**d**) Test data (`test.csv`)

2. The file `feat.csv` contains feature value definitions. This file specifies the names of the features and the dimensions of the analysis data. In this specific example, we have 1,281 rows with 1280 features and one dummy feature.
3. The file `train.csv` contains the training data definition. This file has the same number of rows as `feat.csv`, and each column contains one sample used to train the model.
4. The file `test.csv` contains the analysis data definition, providing the data that are used as classification test data. This file has the same format as the training data file `train.csv`, where each column contains one entire feature vector of the test dataset.

9.4.2   Setup of YGAP Console

After specifying the input data CSV files, we open the YGAP console window. As the first step, we create a new working set by selecting the item **WORKING SET** from the main menu, where we can specify the name of the working set. This automatically creates the files README, `bam.py`, and `cfg.json` in the working

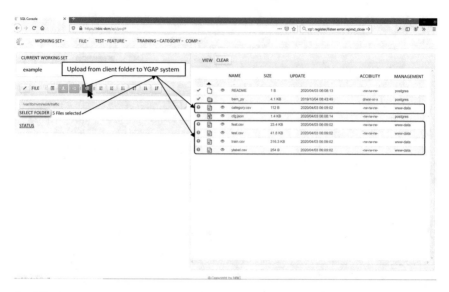

Fig. 9.7 Yuragi Learning General-Purpose Data Analysis Platform (YGAP) console window after input files are loaded into the working set

set directory. Next, we select **OPEN** from the same menu and the name of the newly created working set to load it into our console. To set the CSV input files created in the previous step, we need to select **UPLOAD** from the main menu and navigate to the folder containing these CSV files with **SELECT FOLDER**, after which they appear as entries in the working set window (see Fig. 9.7).

9.4.3 Training the Model Using Training Data

After the input files are loaded into the working set, we proceed with training the model. First, we specify the features that we wish to include in our study. This can be achieved by selecting the entry **TRAINING** from the main menu, which opens the training window in the left area of the console. After clicking **FEATURE GROUP**, a list of all features defined in the file feat.csv is presented. The features that we wish to include in our study can be selected from this list. In this example, we wish to include all features; thus, we click on **SELECT ALL** (see Fig. 9.8).

Individual parameters of the Yuragi learning process with the YGAP, such as the sensory uncertainty r or the dynamics uncertainty q, can be modified by clicking **SETTING**. The training process is started by pressing the **GO** button. This takes several seconds to several minutes depending on the complexity of the model. After the training is completed, graphical plots of the training results appear in the right pane of the console (see Fig. 9.9).

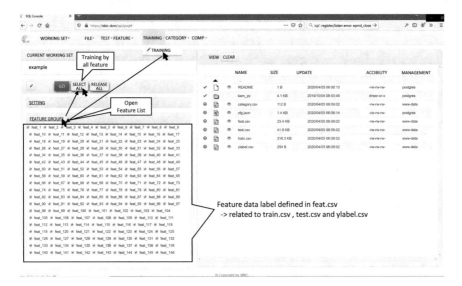

Fig. 9.8 Selecting the features for training and testing in Yuragi Learning General-Purpose Data Analysis Platform (YGAP)

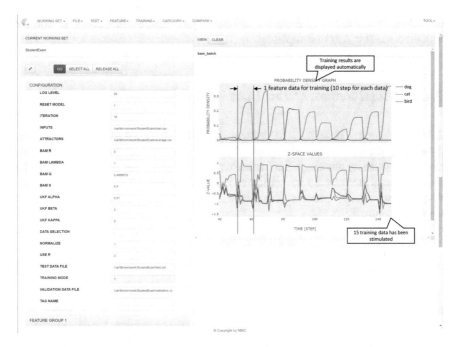

Fig. 9.9 Results after training the model with Yuragi Learning General-Purpose Data Analysis Platform (YGAP)

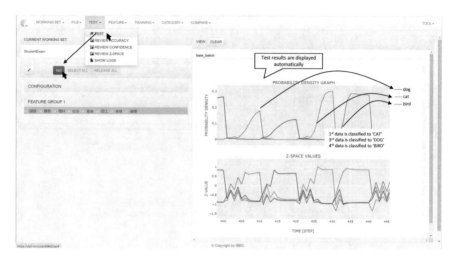

Fig. 9.10 Results of testing data with Yuragi Learning General-Purpose Data Analysis Platform (YGAP)

In this example, 50 samples are trained, and each sample has 1,280 features extracted from images of cats, dogs, and birds. The feature extraction is performed using the shallow part of SimpleNet.

9.4.4 Classification of Test Data

After the training is completed, we can test the performance of our model by classifying data from the test dataset that are not included during training. This can be achieved by clicking **TEST** from the test submenu. The same settings for the feature values are selected for testing as for training and cannot be changed at this stage. After testing is completed, another graphical window displays the results of testing (see Fig. 9.10).

9.5 Conclusion

In this chapter, we consider the cooperation between neuroscience and information science and discuss the development of a new computing system based on research on human cognition. Existing computing systems have evolved with the development of semiconductor technology, steadily increasing the number of computational resources. However, the development of computing systems inspired by the human brain would change the focus of improvement from quantity to quality. In the future, the interaction between the two areas of science will accelerate the development of both. In this chapter, we present the implementation of our YGAP system, which uses Yuragi learning for classification problems. Our proposed platform has the potential to contribute to research beyond the fields of information science and brain science and solve more complex problems that will arise in the near future.

References

1. Bitzer, S., Bruineberg, J., Kiebel, S.J.: A Bayesian attractor model for perceptual decision making. PLoS Comput. Biol. **11**(8), e1004442 (2015)
2. Giese, M., Poggio, T.: Neural mechanisms for the recognition of biological motion. Nat. Rev. Neurosci. **4**, 179–192 (2003)
3. Inagaki, M., Fujita, I.: Reference frames for spatial frequency in face representation differ in the temporal visual cortex and amygdala. J. Neurosci. **31**, 10371–10379 (2011)
4. LeCun, Y., Bengio, Y., Hinton, G.: Deep learning. Nature **521**, 436–444 (2015)
5. Nishihara, K., Taya, N., Kanoh, T.: A consideration of realizing the brain inspired computer. In: Proceedings of the Ninth EAI International Conference on Bio-Inspired Information and Communications Technologies (BICT'15), pp. 495–496 (2015)
6. Nishihara, K., Taya, N., Kanoh, T.: Brain inspired cognitive computing system. In: The Sixth Korea-Japan Joint Workshop on Complex Communications Sciences. Jozankei, Japan (2018)
7. Schaefer, M., Heinze, H.J., Rotte, M.: My third arm shifts in topography of the somatosensory arm. Hum. Brain Mapp. **30**, 1413–1420 (2009)
8. Torii, W., Fujimoto, S., Furukawa, M., Ando, H., Maeda, T.: Techniques to control robot action consisting of multiple segmented motions using recurrent neural network with butterfly structure. In: Proceedings of the International Joint Conference on Biomedical Engineering Systems and Technologies (BIOSTEC 2016), pp. 174–181 (2016)

Chapter 10
Bias-Free Yuragi Learning

Tatsuya Otoshi

Abstract One of the differences between traditional artificial intelligence (AI) and human cognition is the amount of training data required to learn a new category. Conventional AI, such as deep learning, must learn a large amount of data to make decisions; however, humans can make decisions using only several representative examples of data. This ability is essential, especially in situations in which the environment frequently changes, such as the Internet of Things (IoT). With frequent changes in the environment, AI must make decisions based on new situations in real time. Traditionally, several learning methods have been proposed, such as self-training and transfer learning, which assume a small amount of training data. However, when the environment constantly changes and real-time decisions are required, it is necessary to learn and make decisions with a small amount of available data. For this purpose, Yuragi learning, which is a model of decision-making that assumes that obtained information contains uncertainty, is promising due to its noise tolerance. In this chapter, we introduce a system that uses Yuragi learning as a classifier to automatically acquire a new category in new situations. This system detects new categories based on the confidence of existing categories computed by Yuragi learning. Then, the detected categories are added, and new training data are collected in parallel with ordinal classification. The collected data are used as training data for the new category, and additional training is performed to improve the classification accuracy. In this chapter, we apply this classification system to an image classification task for handwritten characters as an example of its application.

10.1 Introduction

Artificial intelligence (AI) technology has experienced remarkable progress in recent years and has a wide range of potential applications [10]. One application

---

T. Otoshi (✉)
Graduate School of Economics, Osaka University, Toyonaka, Osaka, Japan
e-mail: t-otoshi@econ.osaka-u.ac.jp

© Springer Nature Singapore Pte Ltd. 2021
M. Murata, K. Leibnitz (eds.), *Fluctuation-Induced Network Control and Learning*,
https://doi.org/10.1007/978-981-33-4976-6_10

is the Internet of Things (IoT), in which the environment changes frequently due to the movement, addition, or removal of devices. In IoT, it is difficult to collect sufficient training data from the new environment, and the data cannot be handled by traditional AI, which is typically based on large amounts of data. AI also has a large impact on our lives, including our decisions, as it is directly related to our daily activities. However, conventional AI strongly relies on obtained data. With limited training data, bias in the data can cause bias in the judgments of AI.

As a result, it is difficult to adopt traditional AI approaches in situations in which the environment changes frequently and insufficient data about the new environment are available. Two problems, in particular, pose a major challenge: the amount of training data and noise in the data. With respect to the first problem, in a new situation, it takes a certain amount of time to collect data to represent the given situation. However, real-life applications, such as IoT, require AI to make decisions in response to changing circumstances. Therefore, although it is necessary to make decisions based on a small amount of data, data collected over a short period are highly biased, and AI making decisions based on the data will thus be biased as well. With respect to the second problem of noisy data, if the data contain significant noise, it is difficult to classify the category [14]. To respond to new situations in real time, AI must collect the data itself. That is, AI must learn from unlabeled data without manually labeling the data. Because noise makes it difficult to distinguish a new situation from an existing one, it is difficult to label data automatically. The two abovementioned problems are closely related. The lack of data makes it difficult to avoid the influence of noise in the training data, while noise makes it difficult to collect an appropriate amount of training data.

These problems have been addressed in the field of machine learning from various perspectives. One approach is self-training, which is a semi-supervised learning method in which a classifier is trained from a small amount of manually labeled training data and a large amount of unlabeled data [14, 15]. In this method, the classifier is initially trained using a small amount of manually labeled training data. Then, the AI automatically assigns labels to a large amount of unlabeled data using a trained classifier. The classification accuracy can then be improved by relearning the classifier using the data labeled by the classifier itself. Another approach to learning based on a small amount of data is transfer learning [11, 13], in which the classification process performed by a classifier is divided into two stages: feature extraction and classification. In feature extraction, numerical values are output that represent features that characterize the categories of data from non-numerical data, such as images. Because the mapping from non-numerical data to features is generally static, only the process of classification at a later stage is relearned to accommodate the new situation. By avoiding relearning the entire process of classification in this way, the degree of freedom of the classifier decreases, and it is possible to learn even with a small amount of training data. However, even with the abovementioned methods, it is still difficult to learn new categories in noisy environments. Self-training has the potential to lead to biased classification by training with such data, as automatically affixed labels are less reliable in noisy environments. In transfer learning, noise in the training data also affects the accuracy of the classifier. In addition, even if only the classifier is learned, it often includes a considerable number of parameters, and it is difficult to learn only the classifier from a small amount of data obtained in real time.

The noise tolerance of Yuragi learning [4, 12] is a promising feature to solve these challenges. Yuragi learning adopts a model of decision-making [2] that assumes uncertainty in sensing information. In Yuragi learning, the learning parameters are provided as representative values of the observations associated with the alternatives, and a decision is made based on the computed confidence of the alternatives. This signifies that the model is in principle capable of learning even with one-shot data obtained for a new situation. In other words, Yuragi learning has both noise tolerance and the ability to learn from a small amount of data and can learn new categories in real time and make decisions in response to changing situations.

In this chapter, we introduce a system that automatically learns new categories using Yuragi learning, which is a decision-making model that operates as a classifier and is extended to add categories according to the given situation. When using Yuragi learning as a classifier, low confidence is computed for all existing categories when a new category emerges. When low confidence is detected, the system starts to add new categories to the current classifier using the observed data as a representative value. Because the classifier at this time has inadequate and highly biased initial data, we collect training data in parallel with ordinal classification. Then, using the collected data, additional training is performed to improve the accuracy of the classification of new categories.

The remainder of this chapter is organized as follows. Section 10.2 introduces the system components for using Yuragi learning, which is a decision-making model originally, as a classifier. Section 10.3 describes the overall process of the system in detecting and adding new categories, automatically collecting training data, and improving classification accuracy through additional learning. Section 10.4 describes the process of adding new categories with numerical examples, and Sect. 10.5 presents the results of applying the system to an image recognition task for handwritten characters. Section 10.6 provides a summary of the chapter.

10.2 Classification System with Yuragi Learning

This section describes the configuration of a classifier system that classifies categories using Yuragi learning. The classification system consists of two components: a feature extractor and a classifier. The feature extractor extracts a feature, which is input to the classifier, from raw input data, such as image data. Based on the extracted features, the classifier determines the category to which the input data belong. The following subsections describe each component in detail.

10.2.1   Feature Extractor for Preprocessing of Classification

The feature extractor maps non-numerical data, such as images, text, and sound [1, 3, 16], to a numerical vector. The distance of the numerical vector represents the similarity of the original data. This feature representation helps the classifier to classify based on the difference in the intrinsic category of the data rather than the difference in the format of the raw data. For example, for an image, by extracting features of the image that do not depend on the parallel movement in the image, an object whose position is shifted can be easily classified as the same object. For text, by numerically expressing the similarity relation of words, it is possible to classify a sentence by focusing on its meaning regardless of the difference in words used. Such feature extraction is also commonly used in conventional classification in machine learning. Figure 10.1 presents an example of feature extraction of handwritten characters. Each image of a character is mapped into a feature space in which the same characters are mapped into a certain area.

Deep learning is often used as a feature extractor. In a multilayer neural network model, such as deep learning, the feature extraction process is included internally and can be partially extracted to construct a feature extractor. The configuration of deep learning is roughly divided into an input layer, an intermediate hidden layer, and an output layer. The data provided to the input layer are processed in the intermediate hidden layer, and the category is identified in the output layer. Each layer maps the output vector of the previous layer to another vector space, and the output of the intermediate hidden layer can be regarded as the extracted features of the data for the final classification. Therefore, the feature extractor can be constructed by extracting the neural network from the input layer to the intermediate hidden layer. Feature extractors created in this manner are commonly applied to

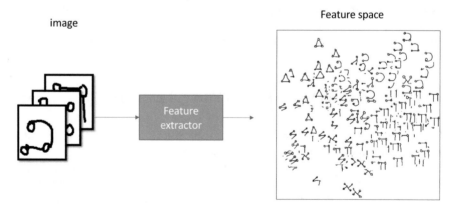

Fig. 10.1 Feature extraction: handwritten characters on the left side are mapped to the feature space on the right side

various types of input, such as words (word2vec), images (image2vec), and sounds (sound2vec) [16]. A neural network is also used as the feature extractor here.

Following [3], we use *triplet loss* as the loss function to learn the neural network for the feature extractor. The triplet loss is defined by the following equation:

$$L = \sum_i \left[\|f(a_i) - f(p_i)\|^2 - \|f(a_i) - f(n_i)\|^2 + \alpha \right]^+,$$

where (a_i, p_i, n_i) represents the triplet of training data, $f(x)$ represents the feature mapping by the neural network, $[x]^+ = \max(0, x)$, and α is a parameter representing how far apart the different categories would be placed in the feature space. The triplet (a_i, p_i, n_i) is selected from the training data so that a_i and p_i are the same categories, while a_i and n_i are different categories. Thus, the triplet loss is minimized when the distance between a pair from different categories is sufficiently larger than the distance between a pair from the same category. That is, the feature extractor learning with triplet loss maps the data into the feature space so that similar inputs are close to one another, while different inputs are far apart.

In a neural network, simultaneous optimization of a feature extractor and a classifier can be performed by combining them into one neural network model. However, the neural network constructed by this combination has many parameters, and a large amount of training data and learning time is required to learn these parameters. Because the characteristics of input data, such as images, are considered to be independent of a particular classification problem, the same feature extractor is often reused for other classification problems. In particular, in a situation in which a new category is acquired, the relationship between the input data and features does not change. Thus, it is preferable to reuse the feature extractor without modifying it and updating only the classifier.

10.2.2 Yuragi Learning as a Classifier

After a feature is extracted from the raw data, the classifier determines the category that the feature represents. That is, the classifier is the mapping from the feature vector to the known category list. It should be noted that grouping the inputs without the category list is called *clustering* [17], which is outside of the scope of this chapter.

In this chapter, we use Yuragi learning as the classifier, which is based on the Bayesian attractor model that formulates the cognitive decision-making process in the brain [2]. In Yuragi learning, the decision-making process is modeled as accumulating confidence with sequential observations. Yuragi learning has several choices called attractors, and it estimates the confidence of each choice in terms of its suitability for the current situation. If the confidence of a choice exceeds a certain threshold, Yuragi learning selects this choice.

We denote the *decision state* at time t as z_t and the state corresponding to S alternatives as $\phi_1, \cdots, \phi_S$. That is, the ith alternative is selected when the decision state $z_t = \phi_i$. Usually, ϕ_i is a one-hot vector whose ith element is 1 and whose other elements are -1.

10.2.3 Yuragi Learning: State Update

When a new observation value x_t is obtained, Yuragi learning calculates the posterior probability distribution $P(z_t|x_t)$ by Bayesian inference. The generative model for the Bayesian inference is as follows:

$$z_t - z_{t-\Delta t} = \Delta t f(z_{t-\Delta t}) + \sqrt{\Delta t} w_t \tag{10.1}$$

$$x_t = M\sigma(z_t) + v_t, \tag{10.2}$$

where f represents the Hopfield dynamics that has S fixed points (or attractors) $\phi_1, \cdots, \phi_S$ presenting the S alternatives. This generative model signifies that when the decision state approaches one of the attractors ϕ_i, the selected attractor is suitable for the observed value x_t. $M = [\mu^{(1)}, \cdots, \mu^{(S)}]$ is a matrix that aligns $\mu^{(i)}$, which is a representative of the observation value for the situation in which the ith alternative should be chosen. σ is a sigmoid function whose image is $[0, 1]$. In this chapter, the function is applied element-wise to the operand if the function is displayed in bold. w_t and v_t are noise terms that follow a Gaussian distribution: $w_t \sim N(0, \frac{q^2}{\Delta t} I)$, $v_t \sim N(0, r^2 I)$. q denotes the *dynamic uncertainty*, which represents the tendency to switch the decision between alternatives, and r denotes the *sensory uncertainty*, which represents that the size of the error is included in the observation value.

10.2.4 Yuragi Learning: Decision-Making

The posterior distribution provides the confidence of an alternative, that is, how suitable it is for the current situation. In Yuragi learning, the decision is determined when the confidence exceeds a certain threshold.

Yuragi learning selects the ith alternative when the following condition is satisfied:

$$P(z_t = \phi_i|x_t) \geq \lambda, \tag{10.3}$$

where λ is the threshold, and $P(z_t = \phi_i|x_t)$ is the confidence of the ith alternative.

10.3   New Category Acquisition in Yuragi Learning

In the original Yuragi learning, the time series of the observation data is input, and each observation data point is recognized as belonging to one of the known categories. Given an observation belonging to an unknown category as input, Yuragi learning calculates the confidence level of each known category to which the observation may belong. Then, it is determined whether the confidence level for a known category similar to the input is high or the confidence level for classifying any known category is low. Thus, there is a bias in Yuragi learning, whereby the classification results differ depending on which category is initially provided as a known category. However, when we observe an unknown object, we recognize it as belonging to a new category, and when we observe the same object again later, we can identify it as belonging to the same category. This ability plays an important role in making appropriate decisions in a changing environment. In IoT applications, AI should have this ability to make decisions in changing situations in real time. Therefore, we consider the expansion of Yuragi learning to have the ability to acquire and classify new categories.

For classification of a new category, the greatest challenge is that AI does not have a large amount of data for the new category. For instance, when data for the new category are observed for the first time, there is only one data point. In an extreme case, it is necessary to acquire the new category from only this point and be able to recognize it when data belonging to the same category are observed again later. Humans can identify any animal from a single photograph and use their abilities to make various judgments daily. Even if a category is initially acquired from one observation data point, knowledge can be corrected through the experience of acquiring additional observation data, and it becomes possible to classify the new category more accurately. Inspired by this human ability, we propose a learning method of Yuragi learning capable of acquiring and modifying a new category online. In the extended Yuragi learning, the classifier is developed by sequentially acquiring a new category while modifying it from a small amount of data that are sequentially obtained.

This chapter proposes a method to add and modify categories online by automatically collecting training data for new categories based on the classified results and accumulating the data as experience in parallel with Yuragi learning classification. This method consists of four processes as follows:

Step 1.   Detect input data belonging to an unknown category based on the confidence level of Yuragi learning.

Step 2.   Calculate the provisional category representative values from a small amount of observed data and add them to Yuragi learning as new categories.

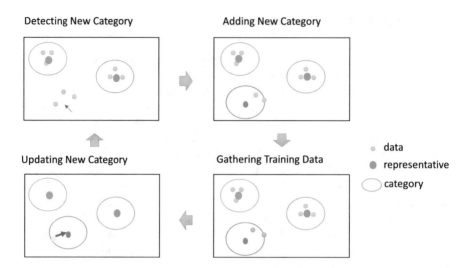

Fig. 10.2 Process of new category acquisition

Step 3.   In parallel with the classification of the category, perform labeling using the confidence of the classification into the new category, and accumulate the label as training data.
Step 4.   Recalculate the category representative value from the accumulated training data to relearn the category.

Figure 10.2 illustrates the overall process. The remainder of this section describes each process in detail.

10.3.1 Detecting New Category

The first step in acquiring a new category is to detect data emerging from an unknown category that do not belong to an existing category. In a subsequent process, a new category is added to Yuragi learning so that the unknown category can be newly classified. In particular, Yuragi learning assumes that observation data are input along with a time series and detects unknown categories in parallel with classification into known categories.

One way to detect new categories is to use the confidence level that Yuragi learning calculates sequentially for incoming observation data. In the classification process, when data belonging to an unknown category arrive, Yuragi learning outputs low confidence for all known categories. Therefore, it is possible to determine that the observed value belongs to the unknown category when the confidence for all known categories falls below a certain threshold.

More precisely, data point x_t is regarded as being from an unknown category when

$$\forall i, \, P(z_t = \phi_i | x_t) \leq \lambda_{new}, \qquad (10.4)$$

where λ_{new} is the threshold for judging that the confidence is low. We denote this new data point x_t satisfying Eq. (10.4) as x_{new}.

A single observed value can be used to determine representative values; however, more accurate representative values can be obtained by collecting a certain number of observed values and detecting the unknown category data from them. However, simply because two data points are in an unknown category, it does not imply that they are in the same category. For example, if dogs and cats are existing categories, both mice and rabbits are in unknown categories; however, they are not in the same category. Nevertheless, it is often possible to assume that temporally continuous inputs are in the same category. For example, in sequential animal photos taken at a zoo, a rabbit begins to appear in the camera and is continuously displayed until it disappears from the frame. As described above, it is thus possible to collectively detect τ pieces of data after the first detection as data of the same unknown category, assuming that continuous inputs at fixed intervals belong to the same category. We denote these τ pieces of data as $x_{new} = x_{new}(1), \cdots, x_{new}(\tau)$.

10.3.2 Adding New Category with Initial Data

When data belonging to an unknown category are detected by the above method, the addition of a new category begins. In Yuragi learning, because a category is expressed by its representative value, a provisional representative value of the category is calculated using the observed data.

In general, representative values of a category represent characteristics that apply to a large amount of data belonging to that category, and a sufficient amount of data is required to calculate an accurate representative value. However, in a real-time scenario, the representative value is calculated from only a small amount of data by the detection of an unknown category. A classifier using this representative value is only a provisional classifier, and in a later step, this provisional classifier is used to collect additional data and improve the classification accuracy.

Specifically, the average value of τ data obtained in the detection of unknown categories is set as a new attractor for Yuragi learning. Therefore, the typical value $\mu^{(new)}$ for a new attractor is as follows:

$$\mu^{(new)} = \frac{1}{\tau} \sum_{i=1}^{\tau} x_{new}(i).$$

Under the assumption that successive input sequences belong to the same category, it is expected that the collected data belong to the same category and the average feature represents a feature common to that category. However, the amount of data is not sufficient, and there is a possibility that a common feature is accidentally extracted depending on the timing of acquiring the data. For example, if a rabbit is detected as a new category when an animal is being photographed with a camera, different characteristics may be obtained depending on the direction of the rabbit photographed at the detection time. Therefore, a provisional classifier using these data tends to be less accurate. In the following two steps, training data are automatically collected using a provisional classifier to accumulate experience and improve the classification accuracy.

10.3.3   Gathering Training Data

The provisional category representative values calculated from a limited amount of data at the time of detection of a new category are likely to be different from the representative values actually common to the category. Therefore, the classifier is improved by collecting data considered to belong to the new category and correcting the representative value. In general, training data are often manually labeled by humans for mechanically recorded data, such as sensor data; however, this is a major challenge in real time due to the time and labor costs involved. Here, the system itself automatically labels the data; thus, the entire process from the detection of a new category to the collection of training data is automatically performed.

After adding a new tentative category, the classifier performs classification of the time series of the input in the same way as the conventional classifier. During the ordinal classification, the obtained data and classification results are accumulated. That is, the new observation value $X = \{x_{t+1}, \cdots, x_{t+T}\}$ obtained after detecting the new category is classified by the provisional classifier, and the classification labels are recorded. Then, a subset $X^{(new)}$ of the observation data classified into the new category is acquired. Here, T is the period of collecting data. Because new categories are added one by one, no new categories are added during this period. Thus, the period should be determined according to the context length in the input time series.

As a result of the above processing, the system acquires the data labeled with the new category from the unlabeled data.

10.3.4   Updating New Category with Gathered Data

The classifier is updated using the new training data automatically collected by the above method. In updating the classifier, the classifier is relearned using the obtained dataset so that each of the known categories and the new category can be differentiated.

A classifier using Yuragi learning is updated by its representative value for each category. Therefore, the representative value of the new category temporarily calculated previously is updated as follows using the new training data $X^{(new)}$:

$$\mu^{(new)} = \frac{1}{|X^{(new)}|} \sum_{x \in X^{(new)}} x,$$

where $|X|$ denotes the number of elements in X. That is, the representative value is set to the average value of the gathered data $X^{(new)}$.

Many machine learning programs, including neural networks, require data even for known categories when new categories are involved in relearning. For this reason, it is necessary to always maintain not only additional training data for the new category but also a large amount of training data used for previous learning. Relearning using such a large amount of data often takes an enormous amount of time. In contrast, in Yuragi learning, relearning can be performed for each category by updating the representative value of each category. In addition, the learning can be performed with a lightweight calculation of the average value of the feature quantity.

By completing this relearning step, a series of operations for adding categories is completed, and the state is shifted to the state of detecting the next unknown category. Then, by repeating the above steps, the system can learn new categories one by one.

10.4   Numerical Simulation

Section 10.3 introduces a method that automatically collects data and learns new categories when an unknown category is detected. The method is described on the assumption that Yuragi learning is used as a classifier; however, similar processing can also be performed using another classifier, such as a neural network. However, Yuragi learning learns by directly updating the representative value, whereas a classifier, such as a neural network, must relearn many parameters contributing to classification. In particular, in the context of learning a new category that occurs during classification, the amount of data that can be used to learn that category is limited, and relearning is difficult with conventional classifiers. Here, a comparison between using Yuragi learning and a neural network as a classifier when adding a new category indicates that Yuragi learning can be learned even with a small amount of training data. To examine the respective features, the operation of the two approaches is compared using a simple numerical simulation. Comparisons in specific applications using image recognition as an example are provided in the next section.

10.4.1 Simulation Scenario

As the simplest scenario for adding a new category, we assume that the system learns one new category from the first two known categories. Initially, the classifier has two categories and classifies a given input as belonging to either category or neither. By presenting data belonging to an unknown category to the classifier, detection of the new classification is initiated according to Step 1 of the new category addition method. The entire process of acquiring new categories depends on which data are detected as a new category by the classifier. However, because we are interested in the amount of data required to acquire a new category, we begin with a state in which one data point is detected to belong to an unknown category for any classifier. Thereafter, the learning of one new category is completed by creating a provisional classifier, automatically collecting training data, and modifying the classifier according to the procedure of the new category addition technique. When the learning of the category is completed, the classification accuracy is measured to examine the differences in performance between the classifiers.

The details of the process are as follows:

1. The classifier is initialized to distinguish the two categories whose mean values are μ_1 and μ_2.
2. New data $x_{new} \sim N(\mu_3, \sigma^2)$ are given to the system, and the system adds the new category.
3. The system collects the training data for the new category with the N data randomly sampled from $N(\mu_i, \sigma^2)$.
4. The system updates the classifier with the collected training data.
5. The classification accuracy is measured with test data sampled from $N(\mu_i, \sigma^2)$,

where $\mu_1 = (1, 0, 0)$, $\mu_2 = (0, 1, 0)$, $\mu_3 = (0, 0, 1)$, and $\sigma = 0.5$.

We change the amount of unlabeled data N to investigate the relationship between the accuracy and the data size. In this numerical simulation, we do not use a feature extractor, as the data are a one-hot vector that already displays the features. For the Yuragi learning parameters, we use the following values: $q = 0.5$ and $r = \sigma$.

10.4.2 Accuracy and Sensitivity

To evaluate whether the acquired new categories are correctly classified, test data generated from the existing categories and added categories are used to confirm that they are classified into the correct categories. Accuracy and sensitivity are used to evaluate the new classification. Accuracy represents the percentage of all test data that are correctly classified and is high when both new and existing categories are correctly classified; however, new and existing categories generally interfere with each other. For example, if a classifier remembers wolves as a new category, data that would have been previously classified as a dog can now be classified as a wolf. When a new category is added, it should be correctly classified without disturbing the correct answers for existing categories. Therefore, we use sensitivity for each category, which evaluates whether each category is correctly classified.

Each metric is defined as follows [6]:

$$\text{accuracy} = \frac{A}{B} \tag{10.5}$$

$$\text{sensitivity}_c = \frac{A_c}{B_c}, \tag{10.6}$$

where A is the number of collect classifications, B is the amount of all test data, A_c is the number of true positives for the category c, and B_c is the amount of test data from the category c.

10.4.3 Using a Neural Network as Classifier

Although this chapter is based on the assumption that Yuragi learning is used as the classifier, our new category acquisition method can also use conventional classifiers, such as neural networks. As a classifier, Yuragi learning can learn based on representative values; therefore, it is a powerful option for classifying new categories when only several data samples are available. To verify the effectiveness of different classifiers, we compare the use of Yuragi learning and a neural network as classifiers.

Unlike Yuragi learning, which adds only representative values of new categories, neural networks, which learn new categories by updating the classifier parameters, must relearn the entire category, including existing categories. Therefore, when a neural network is used, training data for all categories are retained. In addition, when learning is performed, all data, for not only the new category but also existing categories, are used. In contrast, Yuragi learning does not require updating the representative values of existing categories or retaining the training data in the learning process. This time, the model of the neural network consists of three fully connected layers of 3, 64, and 3 neurons.

10.4.4 Results

In our proposed category addition method, training data for a new category are automatically acquired from unlabeled data; however, the amount of acquired training data differs depending on the amount of data to be given at that time, and the classification accuracy changes accordingly. Therefore, the relationship between the classification accuracy and the amount of data to be given is examined while changing the amount of unlabeled data. Figure 10.3 presents the classification

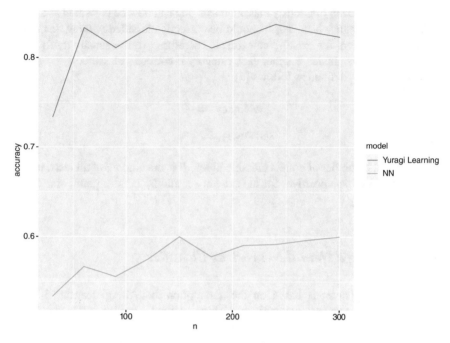

Fig. 10.3 Accuracy of classification when learning a different amount of unlabeled data using Yuragi learning or a neural network (NN) as a classifier. The x-axis (n) represents the amount of unlabeled data

accuracy achieved when a different amount of unlabeled data is provided for both the Yuragi learning classifier and the neural network classifier. Because the data are generated randomly, the accuracy is slightly improved when there is a large amount of data that are easy to classify. Thus, the graph contains small fluctuations in accuracy. However, we focus on the general tendency of the graph.

Figure 10.3 indicates that the classification accuracy when using Yuragi learning is higher than that when using a neural network for any amount of data. This is due to the difference in the way the classifier learns new categories with Yuragi learning and a neural network. In Yuragi learning, new categories are learned using newly obtained training data as representative values. Neural networks, in contrast, learn to classify existing and new categories by modifying the weights of a very large number of neurons. In general, it is difficult to estimate a large number of parameters from a small amount of data. Therefore, a neural network cannot obtain the data necessary for learning when the amount of training data for the new category is limited; as a result, the neural network displays low accuracy. In addition, when the amount of unlabeled data to be given is increased, the classification accuracy of Yuragi learning is greatly improved, particularly when the amount of unlabeled data increases from 30 to 60. This is because with Yuragi learning, the provisional

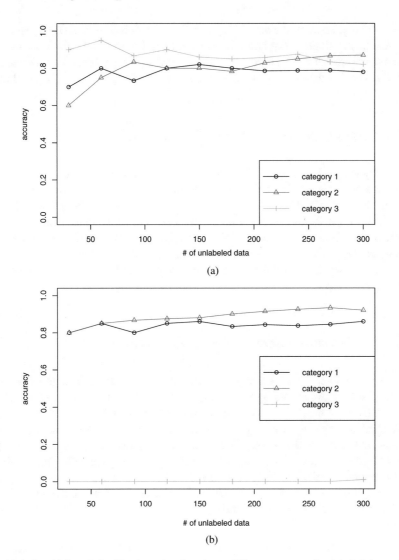

Fig. 10.4 Sensitivity of classification when learning a different amount of unlabeled data using Yuragi learning and a neural network as a classifier. (**a**) Yuragi learning. (**b**) Neural network

classifier is able to correctly collect data belonging to the new category by collecting training data.

Figure 10.4 presents the sensitivity of each category to further characterize the classification of new categories in Yuragi learning and neural networks. The figure indicates that Yuragi learning is highly sensitive to all categories, whereas neural networks are less sensitive to new categories. As described above, a neural network requires learning a large number of parameters; however, in classifying a new category, learning the new category fails because the amount of data available for learning is limited. Furthermore, when a neural network classifier is used as a provisional classifier, it is difficult to collect test data from unlabeled data; thus, even if the amount of data to be given is increased, a new category cannot be obtained.

10.5 Handwritten Character Recognition

To verify the performance of the new category acquisition method, an experiment was conducted using image recognition as an example, which is one of the main applications of AI. Here, the recognition accuracy was evaluated for a scenario in which image data of handwritten characters were sequentially supplied to the system and the system automatically acquired unknown handwritten characters from the image data. The image data of handwritten characters to be supplied to the system used the Omniglot dataset [8], which included character images from various languages.

10.5.1 Evaluation Scenario with Handwritten Character

As in the numerical simulation in Sect. 10.4, we start with a classifier that distinguishes between two types of handwritten characters and acquires the ability to classify new characters. The main difference is that it repeats learning new characters one by one, assuming developmental learning. In other words, a scenario is assumed in which a system that can classify only two characters automatically performs the classification of new characters by providing character images. Here, it is assumed that the detection of an unknown category is performed correctly, and attention is paid to the steps in which the training data are automatically collected by the provisional classifier and the learning proceeds. The detailed process is as follows:

1. The classifier is initialized to distinguish the initial two characters from the Omniglot dataset.
2. A new character is provided to the system, and the system adds a new category.
3. The system collects training data for the new category with unlabeled character images that are sampled from known characters and the new character.
4. The system updates the classifier with the collected training data.
5. The accuracy of classification is measured with test data sampled from known characters and the new character.
6. Steps 2–5 are repeated after the new character is added to the known characters.

10.5.2   Using a Convolutional Neural Network as a Feature Extractor

When classifying an image, it is common to extract features of the image and use them as input to a classifier rather than directly using an image storage format, such as a bitmap, as an input. By extracting the feature quantity, classification based on the characteristics of the category can be performed without depending on the translation of the image. We used a convolutional neural network (CNN) [5, 7, 9] as a feature extractor commonly used in image recognition. The CNN had a simple structure in which a convolutional layer and a pooling layer were repeated twice, and two fully connected layers were applied. The filter size of the convolutional layer and pooling layer was 3×3, and the number of filters of the convolutional layer was 32 and 64. In addition, the fully connected layers at the latter stage set the number of neurons to 64 and 32, and the output consisted of 32-dimensional feature values. The output feature became the input of the classifier. For the classifier, before conducting the image recognition experiment, learning was performed using handwritten images that were not used in the experiment. At this time, the CNN was trained to minimize the triplet loss. By this learning, a feature amount extractor was obtained in which the same character was mapped to a close position on the feature amount space.

10.5.3   Using Neural Network as Classifier

Similar to the numerical simulation in Sect. 10.4, a comparison was performed using a neural network as a classifier. The classifier consisted of three layers: an input layer, an intermediate hidden layer, and an output layer. The input layer had the same 32 dimensions as the feature quantity, the intermediate hidden layer had 64 dimensions, and the output layer had the same dimensions as the number of categories acquired thus far. The learning of the neural network was performed only for the classifier using the training data of the existing category and newly obtained category. At this time, similarly to Yuragi learning, the neural network for feature quantity extraction was fixed and learned. That is, transfer learning [11] was performed for the neural network corresponding to the classifier.

10.5.4   Results

As the number of categories increases, the probability of a classification being correct by chance decreases, as does the accuracy of the classifier. In particular, when new categories are repeatedly learned based on limited data, the accumulation of categories with low classification accuracy reduces the overall accuracy. However,

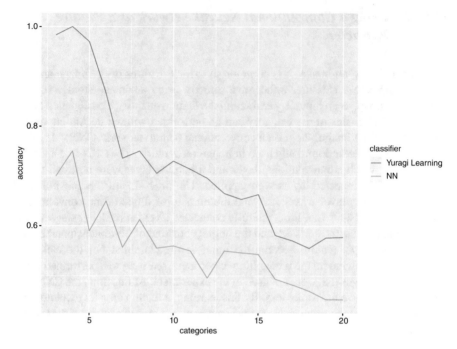

Fig. 10.5 Accuracy of classification after learning the multiple categories using Yuragi learning or a neural network as a classifier

in Yuragi learning, because learning can be performed using only the representative values of the categories, the accuracy of new categories is high, and the degradation in accuracy can be suppressed even if the number of categories increases. Therefore, we investigated the relationship between the classification accuracy and the number of categories. Figure 10.5 illustrates the overall change in classification accuracy as the classifier learned new characters one by one. This figure presents a comparison of the results when Yuragi learning was used as a classifier and when a neural network was used.

Figure 10.5 demonstrates that the classification accuracy decreased with the number of categories, as described above. Moreover, in any process of adding categories, the classification accuracy was higher when Yuragi learning was used as the classifier than when a neural network was used. This is because in Yuragi learning, a new category can be learned by setting a representative value, whereas in a neural network, setting a large number of parameters cannot be performed correctly with limited data. In particular, when using Yuragi learning, the classification accuracy was approximately 1 until the number of categories increased to 5. In contrast, when a neural network was used and the number of categories was 5, the classification accuracy was approximately 0.6. As described above, when Yuragi learning is used, the classification accuracy can be maintained to some extent even if new categories are repeatedly added. However, even when using Yuragi learning, because the accuracy decreases as the number of classifications increases, in practice, it is necessary to present a category with insufficient accuracy when a decrease in accuracy is detected and manually provide accurate training data for that category. Even if this manual process is introduced, the labor will be reduced when Yuragi learning is used because the accuracy remains high, while the number of new categories is small.

10.6 Summary

In this chapter, we introduced a system that automatically collects training data from unlabeled data using Yuragi learning to avoid the bias of a small amount of data. The system uses the confidence of Yuragi learning to detect unknown categories. In addition, it uses detected data to create a provisional classifier. Using this provisional classifier, the system detects data belonging to a new category in parallel with ordinal classification and automatically collects these data as training data. The classification accuracy can be improved by retraining the classifier after collecting additional training data. By repeating this procedure, the system sequentially acquires new categories.

The behavior of the above system was evaluated through numerical simulation. It was demonstrated that a new category could be classified when Yuragi learning was used as a classifier; however, the new category could not be classified when a neural network was used because training data were limited. We also conducted an experiment to learn new categories repeatedly using handwritten character data. The results indicated that when Yuragi learning was used as a classifier, high classification accuracy was maintained even when the number of categories increased in the learning process. In particular, when the learning loop was repeated three times, the classification accuracy was 0.6 when a neural network was used as a classifier, whereas the classification accuracy was almost 1 when Yuragi learning was used as a classifier. In future work, the flexible management of categories by deleting, integrating, and separating categories should be considered.

References

1. Ahmed, E., Jones, M., Marks, T.K.: An improved deep learning architecture for person re-identification. In: Proceedings of the IEEE Conference on Computer Vision and Pattern Recognition, pp. 3908–3916 (2015)
2. Bitzer, S., Bruineberg, J., Kiebel, S.J.: A Bayesian attractor model for perceptual decision making. PLoS Comput. Biol. **11**(8), e1004442 (2015)
3. Hermans, A., Beyer, L., Leibe, B.: In defense of the triplet loss for person re-identification. arXiv preprint arXiv:1703.07737 (2017)

4. Iwamoto, M., Otoshi, T., Kominami, D., Murata, M.: Rate adaptation with Bayesian attractor model for MPEG-DASH. In: 2019 IEEE Ninth Annual Computing and Communication Workshop and Conference (CCWC), pp. 0859–0865. IEEE, Piscataway (2019)

5. Koch, G., Zemel, R., Salakhutdinov, R.: Siamese neural networks for one-shot image recognition. In: ICML Deep Learning Workshop, vol. 2 (2015)

6. Kotsiantis, S., Pierrakeas, C., Pintelas, P.: Predicting students' performance in distance learning using machine learning techniques. Appl. Artif. Intell. **18**(5), 411–426 (2004)

7. Krizhevsky, A., Sutskever, I., Hinton, G.E.: ImageNet classification with deep convolutional neural networks. In: Advances in Neural Information Processing Systems, pp. 1097–1105 (2012)

8. Lake, B.M., Salakhutdinov, R., Tenenbaum, J.B.: Human-level concept learning through probabilistic program induction. Science **350**(6266), 1332–1338 (2015)

9. Lawrence, S., Giles, C.L., Tsoi, A.C., Back, A.D.: Face recognition: a convolutional neural-network approach. IEEE Trans. Neural Netw. **8**(1), 98–113 (1997)

10. Makridakis, S.: The forthcoming artificial intelligence (AI) revolution: its impact on society and firms. Futures **90**, 46–60 (2017)

11. Ng, H.W., Nguyen, V.D., Vonikakis, V., Winkler, S.: Deep learning for emotion recognition on small datasets using transfer learning. In: Proceedings of the 2015 ACM on International Conference on Multimodal Interaction, pp. 443–449. ACM, New York (2015)

12. Ohba, T., Arakawa, S., Murata, M.: A Bayesian-based approach for virtual network reconfiguration in elastic optical path networks. In: Optical Fiber Communication Conference, pp. Th1J–7. Optical Society of America (2017)

13. Torrey, L., Shavlik, J.: Transfer learning. In: Handbook of Research on Machine Learning Applications and Trends: Algorithms, Methods, and Techniques, pp. 242–264. IGI Global, Pennsylvania (2010)

14. Triguero, I., Sáez, J.A., Luengo, J., García, S., Herrera, F.: On the characterization of noise filters for self-training semi-supervised in nearest neighbor classification. Neurocomputing **132**, 30–41 (2014)

15. Veselỳ, K., Hannemann, M., Burget, L.: Semi-supervised training of deep neural networks. In: 2013 IEEE Workshop on Automatic Speech Recognition and Understanding, pp. 267–272. IEEE, Piscataway (2013)

16. Wang, S., Zhang, J., Zong, C.: Associative multichannel autoencoder for multimodal word representation. In: Proceedings of the 2018 Conference on Empirical Methods in Natural Language Processing, pp. 115–124 (2018)

17. Xie, J., Girshick, R., Farhadi, A.: Unsupervised deep embedding for clustering analysis. In: International Conference on Machine Learning, pp. 478–487 (2016)

Index

© Springer Nature Singapore Pte Ltd. 2021
M. Murata, K. Leibnitz (eds.), *Fluctuation-Induced Network Control and Learning*,
https://doi.org/10.1007/978-981-33-4976-6

Printed in the United States
by Baker & Taylor Publisher Services